Ecologics

Ecologics

Wind and Power in the Anthropocene   **Cymene Howe**

Duke University Press   *Durham and London*   2019

© 2019 DUKE UNIVERSITY PRESS ALL RIGHTS RESERVED
PRINTED AND BOUND BY CPI GROUP (UK) LTD, CROYDON, CR0 4YY
DESIGNED BY COURTNEY LEIGH BAKER AND TYPESET IN MINION PRO
AND FUTURA STANDARD BY WESTCHESTER PUBLISHING SERVICES

Library of Congress Cataloging-in-Publication Data
Names: Howe, Cymene, author.
Title: Ecologics : wind and power in the Anthropocene / Cymene Howe.
Other titles: Wind and power in the Anthropocene
Description: Durham : Duke University Press, 2019. | Includes bibliographical
references and index.
Identifiers: LCCN 2018050150 (print)
LCCN 2019000665 (ebook)
ISBN 9781478004400 (ebook)
ISBN 9781478003199 (hardcover : alk. paper)
ISBN 9781478003854 (pbk. : alk. paper)
Subjects: LCSH: Wind power—Research—Mexico—Tehuantepec,
Isthmus of. | Renewable energy sources—Mexico—Tehuantepec,
Isthmus of. | Renewable energy sources—Political aspects. |
Electric power production—Mexico—Tehuantepec, Isthmus of. |
Energy industries—Mexico—Tehuantepec, Isthmus of. | Energy
development—Political aspects. | Energy policy—International
cooperation. | Geology, Stratigraphic—Anthropocene.
Classification: LCC TJ820 (ebook) | LCC TJ820 .H69 2019 (print) |
DDC 333.9/2097262—dc23
LC record available at https://lccn.loc.gov/2018050150

Cover art: Bat falcon in flight. Photo © Juan Carlos Vindas / Getty Images.

This title is freely available in an open access edition thanks to generous
support from the Fondren Library at Rice University.

For Dominic

Contents

Joint Preface to *Wind and Power in the Anthropocene*

A Dynamic Duo

Welcome to our duograph. You may be entering into the duograph through *Ecologics* or *Energopolitics*, but in each case, we invite you to engage both sides of this work. The duograph is a new and experimental form that needs your active engagement. But what is a duograph? you might rightly ask. A duograph consists of two single-authored ethnographies that draw from a shared fieldwork experience and the same archive of research material. As a textual form, the duograph emerged from our field research (2009–13) on the political and ecological dimensions of wind power development in Mexico's Isthmus of Tehuantepec. The idea evolved partly out of experimental interest and partly out of necessity. The two of us spent many long evenings debating the significance of one aspect or another of the research and gradually found ourselves setting out from the center of the project in different theoretical and thematic directions. The fieldwork itself was a joint enterprise from start to finish; every interview, every meeting, every protest, involved both of us. We originally expected that the writing would follow a similar path toward a coauthored monograph. But while coauthoring offers many opportunities to learn and grow through dialogue, it also involves many compromises and ultimately must resolve in a synthetic voice and direction. We wanted to do this differently.

We eventually realized how important it was to each of us that we be able to tell a different part of the immensely complex story unfolding in the isthmus. Cymene wanted to spotlight the salience of human-nonhuman relations in energy transition while Dominic wished to concentrate on unraveling the political complexity of wind power. We decided to experiment by elaborating our different analytics and interests in companion volumes that are meant to be read together. A working definition of the duograph would be a conversation between researchers that materializes in two texts, which do not require analytic synthesis or consensus. We view the duographic form as a way to produce collaborative scholarship that helps to make visible the multiplicity of stakes and attentions existing within the practice of research collaboration. The observations and arguments found in each of these volumes emerged from close dialogue and are by no means incommensurable, but neither are they serial parts of the same narrative. They speak in parallel, but not always in unison. Characters, dynamics, and events crisscross them, but they are approached through different analytic lenses. We hope that the duograph offers an experimental prototype in collective authorship that may be of value to other collaborators and other projects elsewhere.

Wind Power in Mexico

Our ethnography addresses a central question of our anthropocenic times: How can low-carbon energy transition happen? Or, put differently, What happens in those transitions? Who sets the agenda? Who—human and otherwise—is affected? And what are the political (in the broadest sense of the term) forces that shape the possibilities for low-carbon energy futures?

These questions initially took shape at Busboys & Poets café in Washington, DC, in late 2008 as we prepared for a move to Houston, Texas, a global epicenter of the fossil fuel industry. We considered a number of different fieldsites of renewable energy production that appeared to be poised for rapid development. We looked at the DESERTEC solar project in Morocco and nascent programs of wind development in Venezuela and Brazil among other cases. But the one that attracted and held our attention most strongly was Oaxaca's Isthmus of Tehuantepec.

A gap in the Sierra Madre Mountains creates a barometric pressure differential between the Gulf of Mexico and the Pacific Ocean, forming a wind tunnel in the isthmus where wind speeds regularly flirt with tropical storm strength. The *istmeño* wind is capable of overturning semitrailers with ease, uprooting trees, and stripping the paint off boats. This region—often said to be the least developed in a state that is the second poorest in Mexico—is considered to have among the best resources for terrestrial wind power anywhere in the world. That potential was first tapped in the mid-1990s through government demonstration projects designed to lure transnational investment in renewable energy production. But wind development only really gained attention and momentum during the administration of President Felipe Calderón (2006–12). Although Calderón's administration is better known for its drug war and for ceding sovereignty to cartels and capital, his climate change advocacy transformed Mexico from a pure petrostate into a global leader in low-carbon energy transition. Mexico passed some of the most ambitious, binding clean-energy legislation anywhere in the world, including a legal mandate that 35 percent of electricity be produced from non-fossil-fuel sources by 2024, with 50 percent of that green electricity expected to come from wind power, and with most of that wind power expected to come from the Isthmus of Tehuantepec. Private-Public Partnerships (PPPs) in wind energy development mushroomed rapidly. Between 2008 and 2016 the wind energy infrastructure of the isthmus expanded from two wind parks offering 85 megawatts of production capacity to twenty-nine wind parks with 2,360 megawatts of capacity, a 2,676 percent increase in less than a decade that has made the isthmus the densest concentration of onshore wind parks anywhere in the world. → Why is Mexico still underdeveloped then?

Over the course of sixteen months of field research (in 2009, 2011, and 2012–13), we sought to cast as broad a net as possible and speak with representatives of every group of "stakeholders" in wind development in Mexico. Conversations with community members and corporate executives; federal, state, and local government officials and NGO staff; industry lobbyists and antiwind activists; conservationists and media professionals; indigenous rights advocates, bankers, and federal judges, all provided a meshwork of perspectives, which we traced as we moved between the many communities of the isthmus; to the state capital, Oaxaca City; and finally to the federal capital, Mexico City. In total, we conducted more than three hundred interviews and participated in hundreds of hours of less formal conversations. Working with a team of local researchers, we were able to conduct the first door-to-door survey of reactions to wind development in La Ventosa—one of two isthmus towns that are now nearly completely encircled by wind parks. We sat in on governmental and activist strategy meetings and toured wind parks. We marched, rallied, and stood at the fulcrum of many roadblocks erected by opponents of the wind parks. We witnessed the evolving politics of solidarity between *binnizá* (Zapotec) and *ikojts* (Huave) peoples whose shared resistance to particular forms of energy infrastructure brought them into alliance after hundreds of years of interethnic conflict. We arrived at and left fieldwork as committed advocates for low-carbon energy transition. But our experiences in Mexico taught us that renewable energy can be installed in ways that do little to challenge the extractive logics that have undergirded the mining and fossil fuel industries. Renewable energy matters, but it matters more how it is brought into being and what forms of consultation and cooperation are used. We thus came to doubt that "wind power" has a singular form or meaning. Everywhere in our research, it was a different ensemble of force, matter, and desire; it seemed inherently multiple and turbulent, involving both humans and nonhumans. To capture that multiplicity, we came to think about our object of research as "aeolian politics," borrowing from the Spanish term for electricity derived from wind power, *energía eólica*.

Three case studies of aeolian politics came to absorb us in particular—Mareña Renovables, Yansa-Ixtepec, and La Ventosa—the first is the most complex and is treated at length in the *Ecologics* volume. The other two are highlighted in the *Energopolitics* volume. All three represent distinct configurations of aeolian politics; two can be categorized as cautionary tales of failure and the other as an example of the successful achievement of what for many is the renewable dream come to life. And yet success and failure were always in the eyes of their beholders. In all three studies we have sought to balance the fact of anthropogenic climate change and the need for global decarbonization against the local salience of vulnerable statecraft, demands for indigenous sovereignty, and the other-than-human lives that inhabit the Isthmus of Tehuantepec.

Volumes

ECOLOGICS

Ecologics tells the story of an antidote to the Anthropocene, one that was both a failure and a success. The Mareña Renovables wind park would have been the largest of its kind in all Latin America, and it promised immense reductions in greenhouse gas emissions as well as opportunities for local development. In *Ecologics* we follow the project's aspirational origins as well as the conflicts and ethical breakdowns that would leave it in suspension. Drawing from feminist theory, new materialisms, and more-than-human analytics, this volume of the duograph examines the ways that energy transitions are ambivalent: both anticipatory and unknown, where hope and caution are equally gathered. In the case of Mareña Renovables, distinct imaginaries of environmental care and environmental harm were in conflict, effectively diagnosing the deeply relational qualities of energy and environment. The core argument that *Ecologics* advances is that the contemporary dynamics of energy and environment cannot be captured without understanding how human aspirations for energy articulate with or against nonhuman beings, technomaterial objects, and the geophysical forces that are at the center of wind power and, ultimately, at the heart of the Anthropocene.

The analytic architecture of *Ecologics* is both anticipatory and interruptive, and readers are encouraged to engage with the work in an itinerant and wandering way. Three chapters focus on the case of the Mareña project, tracing its inception and the policy regimes and economic conditions that allowed for its initial development (chapter 2, "Wind Power, Anticipated"), following it through a series of dramatic standoffs and protests against the park's creation among indigenous and mestizo communities in the isthmus (chapter 4, "Wind Power, Interrupted"), and finally witnessing the collapse of the wind power project itself resulting from multiple political, economic, and communicational impasses (chapter 6, "Wind Power, In Suspension"). These chapters are interrupted by others that focus on wind, trucks, and species respectively. The interruptive design is intended to mime the empirical, ethnographic dynamics of the research, where forces (like wind), technomaterial tools (like trucks), and other-than-human beings (creatures of all kinds) came to stall and vex human-designed notions of progress and infrastructural development. In *Ecologics* creatures, materials, and elemental forces are bound up with wind power as an analytic object, and they in turn invite new human responses to the paradoxes we face in a time of climatological uncertainty.

ENERGOPOLITICS

Energopolitics engages the case of Mexican wind power to develop an anthropological theory of political power for use in the Anthropocene anchored by discussions of "capital," "biopower," and Dominic's own neologism, "energopower." At the same time, the volume emphasizes the analytic limitations of these conceptual minima when confronted with the epistemic maxima of a situation of anthropological field research on political power. Those maxima not only exceed the explanatory potential of any given conceptual framework, they also resolutely demand the supplementary analytic work of history and ethnography. Concretely, the volume argues that to understand the contemporary aeolian politics of the Isthmus of Tehuantepec, one needs to understand, among other things, a contested history of land tenure, *caciquismo* (boss politics), and student/teacher/peasant/worker/ fisher opposition movements specific to the region; the phantasmatic status of state sovereignty within Mexican federalism; the clientelist networks and corporatist machinations of the Mexican political parties; the legacies of settler colonialism; a federal government anxious about waning petropower and climate change; and a vulnerable parastatal electricity utility trying to secure its future in an era of "energy reform." These forces are just as critical to Mexico's aeolian politics as the processes and dynamics that are duly captured by concepts such as capital, biopower, and energopower. *Energopolitics* is thus an urgent invitation for Anthropocene political theory to unmake and remake itself through the process of fieldwork and ethnographic reflection.

The invitation unfolds across five ethnographic chapters, each highlighting a different localization of aeolian politics. We begin with the as-yet failed effort to build a community-owned wind park in Ixtepec, then move east to the town of La Ventosa, which is successfully encircled by turbines that were built in the dominant PPP paradigm, yet has also been beset by uncertainty and unrest. We encounter the performative sovereignty of the state government in Oaxaca City as it searches for a means to regulate and profit from wind development and then journey northwest to Mexico City to interview those in government, industry, and finance who firmly believe they are steering the course of wind power in the isthmus. Finally, we return to Juchitán, which is not only the hub of local aeolian politics in the isthmus but also a town whose citizens imagine themselves to be the inheritors of a decades- if not centuries-long tradition of resistance against the Oaxacan and Mexican states. In this way, *Energopolitics* seeks to speak *terroir* to *pouvoir*, highlighting the need to resist anthropocenic universalism by paying attention to the profound locality of powers, agents, and concepts. As Claire Colebrook has argued, recognition of the Anthropocene should mark the "return of difference" that has been long called for in feminist and ecological criticism.

Collaboration in Anthropology

Our duograph belongs to a long history of anthropological collaboration in research and writing. In the early decades of North American and European ethnology, the discipline's close ties to fields like geography and natural history meant that the scientific expedition was an important apparatus of anthropological research practice. In the late nineteenth and early twentieth centuries, projects of linguistic and cultural salvage and analysis remained closely allied with archaeology and museology, which explains how some of the most ambitious and important collaborative anthropological enterprises of the era—Franz Boas's Jesup North Pacific Expedition (1897–1902), for example—were organized principally around building natural history collections. As the twentieth century wore on, an individualistic model of field research came to predominate in American and European anthropology, at least normatively, and was celebrated for the transformative qualities of participant-observational immersion. But one would scarcely have had to scratch the surface of any ethnographer-informant dyad to illuminate the complex webs of social enablement—involving research assistants, translators, laborers, intermediaries, government agents—that made anthropological research in the classic Malinowskian mode possible.

After the Second World War, a new emphasis on interdisciplinary area studies research in the social sciences expanded and intensified anthropology's range of collaborative engagements around the world. Much as expedition-era anthropology was absorbed into colonial and imperial knowledge projects, the area studies era was imbricated with the national and international political dynamics of the Cold War. Governments sought to enroll anthropologists in military and intelligence operations across the world—Project Camelot being one of the most well known. However, anthropology was also broadening its epistemic ambitions and moving from cultural salvage projects toward a grappling with modernity and the complex cultural and social dynamics of cities, nations, and world systems. Interdisciplinary exchanges no doubt served to accelerate this shift. And 1950s enterprises like Cornell's Vicos project in Peru (creating a "laboratory for social change") or the MIT Modjokuto project in Indonesia (which gave Clifford and Hildred Geertz their first fieldwork opportunity) cultivated the kinds of long-term interdisciplinary research networks that influenced graduate training and pedagogy as well.[1]

The postwar period also saw an efflorescence of anthropological research partnerships mediated through marriage and other life partnerships. Margaret Mead and Gregory Bateson are a classic example, Margaret Mead and Ruth Benedict a more elusive but possibly more substantial one. Then came the Geertzes as well as June and Manning Nash, Marilyn and Andrew Strathern, Edith and Victor Turner, and Margery and Eric Wolf, followed later by Barbara and Dennis Tedlock, Michelle and Renato Rosaldo, Sally and Richard Price, and Jean and John Comaroff, among others. Anthropology has seen many couples practice the crafts of research, teaching, and writing under at least a partly shared sense of identity, each navigating its own relational dynamics as well as the dominant masculinist heteronorms of the discipline and the university in the twentieth century.

Reacting to the still broader and more complex scale of post-1980s globalization and its social, economic, and environmental consequences, the twenty-first century has seen renewed interest in collaborative research partnerships. Three that have inspired our duograph in particular have been the Matsutake Worlds Research Group (Anna Tsing, Shiho Satsuka, Miyako Inoue, Michael Hathaway, Lieba Faier, and Timothy Choy), the Ethnographic Terminalia collective (Craig Campbell, Kate Hennessy, Fiona McDonald, Trudi Lynn Smith, and Stephanie Takaragawa), and the Anthropology of the World Trade Organization group (Marc Abélès, Máximo Badaró, Linda Dematteo, Paul Dima Ehongo, Jae Aileen Chung, Cai Hua, George Marcus, Mariella Pandolfi, and Phillip Rousseau).[2] All are multi-institutional and international partnerships that have explored new ways of creating anthropological knowledge by crossing the boundaries between anthropological research practices and the arts.

Collaboration itself is nothing new in anthropology; there is abundant evidence that it has been a productive dimension of anthropological research and writing since the discipline's beginning. Further, intimate research partnerships have long fueled the production of anthropological knowledge. There is doubtless an important book to be written about how the particular qualities, subjectivities, and dynamics of particular collaborations have influenced the kinds of knowing and knowledge that those enterprises generated. But our intervention here is more limited. We have found it striking that the spirit of collaborative research has not always translated well into practices of authorship. Coauthored texts remain the exception rather than the rule in anthropology, even when they derive from jointly undertaken field research.[3] The reasons for this gap are not simple and involve considerations ranging from professional reputation to relational dynamics to institutional audit cultures that seek to impose a mathematics of individual accomplishment and accountability on the sociality of research, analytic, and writing practices. What is striking in our view is that there are relatively few models for collaborative writing beyond the model of the jointly authored single text that synthesizes analytic perspectives under a common "we." This is why we have centered our methodological intervention on the duographic form: we are looking for ways to strike a better balance between individual ideation and expression and collaborative fieldwork and archiving.

An important added benefit of the duograph is that it permits a more extensive analytic division of labor between its volumes, as parallel yet distinct arguments can be developed with respect to the common research archive. In our case, the *Ecologics* volume's close focus on how human energetic and environmental aspirations intersect with other-than-human beings and agencies complements yet also reframes the *Energopolitics* volume's effort to offer a more nuanced and comprehensive set of analytics of (human) political power, and vice versa. If the general premise of the entire research project has been that a certain politics of energy is creating a situation of ecological emergency, then it is fitting, and we might say necessary, to be able to offer detailed conceptual and ethnographic accounts of both sides of the equation—energopolitical and ecological. Had we tried to compact all these storylines into a single, synthetic account, however, we might well have burst its seams or have been forced to simplify matters to the extent that neither side would have received its due. In the duographic form, meanwhile, two volumes working together in the mode of "collaborative analytics" can dive deeply into different dimensions of the research while still providing valuable ethnographic elaboration and conceptual infrastructure for each other.[4]

Your Turn

One of our favorite rationales for the duograph is what is happening right now: you are deciding where to start. True to the lateral media infrastructures and expectations of this era, we aspire to offer a more dialogic, collaborative matrix of encounter with anthropological writing. We have sought the words to write; you now seek the words to read. We have left signposts as to where we think the volumes intersect. But you can explore the duograph as you like, settling into the groove of one narrative or zigzagging between them. Think of it somewhere between a Choose Your Own Adventure book and open-world gameplay. Follow a character, human or otherwise; riddle through the knots and vectors of aeolian politics; get bogged down somewhere, maybe in the politics of land or the meaning of trucks; then zoom back out to think about the Anthropocene. Or perhaps pause for a minute or two to watch the birds and bats and turbines that now populate the istmeño sky.

Cymene Howe and Dominic Boyer

Acknowledgments

We gratefully acknowledge the following persons and institutions whose support made this duograph possible. The preliminary field research for the project was funded by a grant from the Social Sciences Research Institute at Rice University. Rachel Petersen and Briceidee Torres Cantú offered extraordinary research assistance in the early research phase. The main period of field research in 2012–13 was funded by the Cultural Anthropology Program of the Division of Behavioral and Cognitive Sciences at NSF (research grant #1127246). We further thank the Social Sciences Dean's Office and especially our colleagues in Rice Anthropology—Andrea Ballestero, James Faubion, Jeff Fleisher, Nia Georges, Susan McIntosh, and Zoë Wool—for absorbing our share of administrative, advising, and teaching duties to allow us to undertake field research for a full year. We give thanks as well to our dear colleagues in Rice's Center for Energy and Environmental Research in the Human Sciences (CENHS)—Bill Arnold, Gwen Bradford, Joe Campana, Niki Clements, Farès el-Dahdah, Jim Elliott, Melinda Fagan, Randal Hall, Lacy M. Johnson, Richard R. Johnson, Jeff Kripal, Caroline Levander, Elizabeth Long, Carrie Masiello, Tim Morton, Kirsten Ostherr, Albert Pope, Alexander Regier—and the Center for the Study of Women, Gender and Sexuality (CSWGS)—Krista Comer, Rosemary Hennessy, Susan Lurie, Helena Michie, Brian Riedel, Elora Shehabuddin, and Diana Strassmann— for much-appreciated moral and intellectual support during the writing phase of the project.

In the Isthmus of Tehuantepec, our debts are many. We would first of all like to thank the Istmeño binnizá and ikojts communities of Álvaro

Obregón, Ixtepec, Juchitán, La Ventosa, San Dionisio del Mar, and Unión Hidalgo for hosting us during different phases of the field research and for a great many conversations and small acts of kindness that helped this project immeasurably. There were certain individuals in the isthmus who became friends and allies in the course of this research and without whose support this duograph would never have been possible. Our special thanks to Isaul Celaya, Bettina Cruz Velázquez, Daniel González, Mariano López Gómez, Alejandro López López, Melanie McComsey, Sergio Oceransky, Rodrigo Peñaloza, Rusvel Rasgado López, Faustino Romo, Victor Téran, Vicente Vásquez, and all of their families. Our condolences to the family of David Henestrosa, a brave journalist who passed away shortly after our field research was completed.

In Oaxaca City, Mexico City, and Juchitán, there were several organizations whose representatives helped enable particular parts of our research, including AMDEE, APIIDTT, APPJ, CFE, CNTE (Sec. 22), CRE, EDF Energies Nouvelles, Iberdrola Mexico, SEGEGO, SEMARNAT, SENER, UCIZONI, and the Universidad Tecnológica de los Valles Centrales.

We would also like to thank several individuals who helped us navigate the complex world of Mexican wind power. Thanks to Fernando Mimiaga Sosa and his sons for convincing us to make Oaxaca the focus of our research, to Sinaí Casillas Cano for his insights into the political contingencies of Oaxacan state governance of renewable energy development, to Felipe Calderón for answering our questions concerning the importance of wind power during his sexenio, to Laurence Iliff for a wealth of insights into the behind-the-scenes politics of Mexican energy, to Dr. Alejandro Peraza for an invaluable series of contacts among the power brokers of Mexican energy, and to The Banker, who remains anonymous but who gave us critical insights into the financial dimension of wind power development.

Both the project and the duograph have benefited immensely from comments at several meetings of the American Anthropological Association, the Cultures of Energy Symposium Series, and the Petrocultures Research Group. Conversations with colleagues in the following forums were also critical in shaping the thinking that has gone into the manuscripts: the Climate Futures Initiative, Princeton University; Association for Social Anthropology Annual Conference, Durham, England; Earth Itself: Atmospheres Conference, Brown University; Energy Ethics Conference, St. Andrews University; Religion, Environment and Social Conflict in Contemporary Latin America, sponsored by the Luce Foundation and the Center for Latin American and Latino Studies, American University; Hennebach Program in the Humanities,

Colorado School of Mines; Klopsteg Seminar Series in Science and Human Culture, Northwestern University; Colloquium for the Program in Agrarian Studies, Yale University; Rice University's Feminist Research Group and CENHS Social Analytics workshop; Environments and Societies Institute, University of California, Davis; Global Energies Symposium, University of Chicago Center for International Studies; Durham Energy Institute; Yale Climate and Energy Institute; Rice University, Mellon-Sawyer Seminar; Centre for Research on Socio-Cultural Change, Manchester University; After Oil, University of Alberta, Edmonton; Institut für Europaïsche Ethnologie, Humboldt Universität zu Berlin; Electrifying Anthropology Conference, Durham University, sponsored by the Wenner-Gren Foundation; Parrhesia Masterclass, Cambridge University; Alien Energy Symposium, IT University of Copenhagen. Thanks also goes out to friends and colleagues at the Departments of Anthropology at the London School of Economics; Ustinov College, Durham University; University of California, Los Angeles; University of California, Davis; Yale University; and University College London who were kind enough to invite us to share our work-in-progress with them.

Paul Liffman has been immensely helpful throughout the life of this project, offering smart critique and always ready to provide context by drawing on his truly encyclopedic knowledge of Mexico's political, economic, and social worlds. Our collaboration with Edith Barrera at the Universidad del Mar enriched our field experience, and back in the United States we were lucky to work with Fiona McDonald, Trudi Smith, and Craig Campbell, members of the Ethnographic Terminalia Collective, to imagine/incarnate the wind house installation of Aeolian Politics at the Emmanuel Gallery in Denver, Colorado. Lenore Manderson was also kind enough to review that project of imagining ethnography and wind differently. Anna Tsing and John Hartigan provided marvelous feedback on earlier chapters, and in later stages, the duograph benefited greatly from anonymous reviewers' incisive commentaries. We appreciate especially their willingness to think with us on the experimental form of the duograph and for their wise handling of interconnections between the volumes.

Thanks are also owed to the wonderful professionals at the University of Chicago Press, especially Priya Nelson, as well as Stephanie Gomez Menzies, Olivia Polk, Ken Wissoker, Christi Stanforth, and Courtney Baker at Duke University Press.

Introduction

On many afternoons, it is the windiest place on earth.

Carving out the narrowed girth of southern Mexico, the Isthmus of Tehuantepec is home to a anemometric quality that is nearly unmatched. Wind is valuable here, its steady pulse an ideal quotient of kinetic force to turn the blades of turbines that, in turn, make electricity. With this wind development might follow; with this wind new wealth might follow. And these are two of the reasons why the Oaxacan isthmus now represents the densest concentration of wind parks on land anywhere in the world. But in this wind other things are also gathered and captured: birds and turtles, trucks and barricades, dirt and money.

This book began as a way to follow wind, and wind power, as a "salvational object": a social and technical apparatus to mitigate climate change in environmentally precarious times. How wind power was being located—epistemically, infrastructurally, and politically—were my abiding questions at first. However, what was initially an exercise in political-economic reasoning or an accounting of resources and their manipulations, became something more. Across hundreds of conversations and thousands of hours of encounter, it became increasingly clear that energy transition is not the work of people alone. In questions of power, both energetic and political, people's aspirations and their cosmological views are crucial. But it is also the case that human actions can never disclose the full extent of how new energy forms are able to reassemble the lifeworlds of creatures, or how they can shape the potential of inanimate things. Concentrating only on the sociocultural dimensions of energy risks obscuring others, particularly how

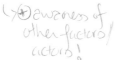

the elemental force of wind, in itself, might become differently. Deeply political projects of renewable energy development and the rise of wind parks have come to occupy the Isthmus of Tehuantepec. Unequivocally. But coincident with this truth are others: biota and stones, machines and infrastructures, dust and air. It was in the wind itself that my attentions first became bent, because when it blows and when its velocity and pressure reach their apex, the wind insists that everything is much more than *anthropos*.

This book follows the wind, but it also describes an antidote to the Anthropocene—the epoch of human imprint upon all earth systems from the geologic to the biotic, from the chemospheric to the hydrological, and from the cryospheric to the atmospheric. As a concept, the Anthropocene hails a particular kind of encounter between deep time and human habit; it is meant to highlight a genealogy of consequences as well as presage precarious futures. Anthropogenic impacts from energy extraction, production, and use have surfaced the reciprocal relationships between excess and deprivation, and they have become harbingers for the unsustainable logics that have driven petromodernity.[1] The material forms and interactions that we call "energy" have always been harvested from what the industrialized world has named "the environment." But if the paradigm of the environment has sought to emphasize interdependencies and mutualities, the human-energy nexus has increasingly come to reveal the corrosive ways that people and energy sources interact. Widespread human demands to have energy at our disposal present a calculus between human aspirations for power, human attempts to manage the climate, and the vital possibilities of all creatures, plants, and beings.[2] Within the parallel worlds of energy and environment, it has become clear that although renewable energy transitions demand the adoption of less catastrophic fuel sources, equally critical is understanding how human energetic desires—for light and heat, for movement and flourishing—either correspond with or deeply disrupt the energetic needs of other biotic life and the systems on which we all depend.[3] Therefore, the argument that I develop throughout this book is that we cannot capture the contemporary dynamics of energy and environment without attending to an array of other-than-human relations including those with nonhuman beings, technomaterial artifacts, infrastructures, and geophysical forces.

By exploring the routes and passages between energies and environments, lives and machines, and the forces that compel them, I also want to create a narrative patterning that is at once anticipatory and interruptive. An anticipatory approach is instructive in times that are marked by ecological discord because it attunes our attention to the subjunctive future of the might be.[4] In anticipation, prognoses float, undetermined but not unknown; questions are raised, but conclusions hang in suspension. Wind also enters here as an interruptive force, awakening air and rousing it from stillness. In following the wind, intermittencies find their way into these pages through things like birds and dust, dead dogs and trash, gusts and stillness. The architecture I have developed across the book is likewise anticipatory and interruptive; chapters oscillate back and forth between the contentious development of the massive Mareña Renovables wind park and the ways in which particular other-than-human forces and entities came to challenge that project.

Parallel to the case of Mareña Renovables, I devote attention to three distinct other-than-human actors in the saga of wind power: wind, trucks, and species. Each of these entities has had a profound role in the development of wind power. Strong and steady wind is, of course, a prerequisite for the development of wind power; it is elemental in the most literal way. But trucks, like wind, are also everywhere within wind power in the isthmus, moving men and materials and operating to create particular political outcomes. And in the places where wind power is being developed, there are also myriad species with those threatened by industrial-scale wind parks appearing in particularly stark forms.

While wind, trucks, and species all hold ethnographic resonance for the case of wind power in Oaxaca, each also provides an analytic for the scalar thinking that the Anthropocene demands. They mark temporal coordinates both past and future. Wind that blew centuries ago was a force that can be said to have partly inaugurated the Anthropocene. It was wind power, after all, that blew colonial exploration and imperial exploitation to the New World. In the mid-twentieth-century "Great Acceleration" of carbon use, trucks served as a mechanism to embody the work of fossil-fueled modernity. And finally, in the precarious future of Anthropocene conditions, there are species—the compendium of all known life hanging in the balance in an unbalanced world. Species includes flora and fauna as well as a future humanity, all of which now face uncertain geoenvironmental risks. Through knowing wind power more closely—in its elemental, technological, and more-than-human forms—my hope is to assemble ecologics differently and to look for a new, turbulent prototype.

How Wind Collapses, in the Future Subjunctive

————

We could begin anywhere on a continuum, tilting toward one position or the other, and find ourselves, ultimately, in a story of utter failure or a tale of extraordinary success. Here are two scenarios of how wind collapses in the future subjunctive.

SCENARIO 1

The Mareña Renovables wind park would have been the largest wind park in all Latin America.

It would have generated almost 400 megawatts of electricity, enough to power more than 600,000 Mexican homes. With the isthmus wind as its resource, the 132 turbines and their generators would have prevented almost 900,000 tons of carbon dioxide emissions every year for at least two, perhaps three, decades. Financed by a consortium of international investment programs—part Japanese, part Australian, and part Dutch—the Mareña park would have put millions of dollars into the hands of environmentally conscious investors, providing capital for sustainable infrastructure projects in the future. The companies that would have purchased the clean power from the park would have benefitted from receiving bonos de carbono (carbon credits), offsetting their carbon footprint, and burnishing their profiles as socially responsible corporate enterprises. In the isthmus, jobs would have multiplied during the construction phase, giving work to unionized laborers from across southern Mexico. Local construction companies would have sold their goods. Once built, the project would have provided jobs in engineering, maintenance, and management. Indigenous communal landholders, ikojts (Huave) and binnizá (Zapotec) people, would have collected millions of pesos for the lease of their land. These funds, in the hands of farmers and fisherfolk, would have been invested in better homes, equipment, and education.[5] More things would have been bought. People would have been healthier and happier, and development would, at last, have arrived in the more remote regions of the isthmus. Roads would have been paved, streetlights would have appeared. Politicians and agencies of government would have been pleased. Mexico would have stood prouder, leading countries of the global South toward greener futures.

The Mareña park, like many of the wind power projects now occupying the isthmus, would have had all of the signatures of success, including

immense amounts of transnational capital and unflinching state sup-
port. It would have been devoted to a new regime of energy that not
only would have empowered Mexico but also would have lived as an
energetic infrastructure to heal the world's wounded climate.[6] But this
would be a failure.

The Mareña Renovables wind park would have been the largest wind
park in all Latin America.

It would have occupied seventy-three acres of territory, its ivory tur-
bines arcing across a sliver of land between the Laguna Superior and
the Laguna Inferior. The territory where it would have been erected
is a biogeographically vulnerable place, a narrow sandbar, or barra.
This sandy stretch of land would have been asked to support 132 tur-
bines, each one reaching 105 meters (thirty-two stories) into the sky
and weighing 285 tons. The 132 towers of steel, many tons of cement,
and 396 blades churning the air might have created quakes in the sand,
sending tremors across the lagoons. Lights atop the turbines would
have burned day and night, and fish, shrimp, and other seaborne life
might have retreated and migrated, leaving local fisherfolk without
their daily catch. The fish might never have returned. And local com-
munities in San Dionisio del Mar, Álvaro Obregón, and Juchitán might
never have fished again. Construction work would have displaced many
tons of mud and earth, and the docking stations where steel and con-
crete would have been off-loaded would have forever changed the
barra. Jobs constructing the park would have gone to outsiders, not to
residents of the region. And the work that would have materialized for
local laborers would have been brutish, short, and poorly paid. For the
lease of their land, some would have become richer while others would
not. Frictions would have endured. Corporate lawyers would have de-
signed the contracts as "evergreen with right to cancel," meaning that
landholders would have indentured their lands for decades. Automatic
renewals on lease contracts would have come to feel very much like
dispossession, or despojo—being robbed of one's land. Indigenous and
campesinos' lands would appear, once again, to be vulnerable to the
whims of the transnationals. Members of the comuna *(communal es-*
tates), or comuneros *(communal landholders), who originally signed*
contracts would feel that they had never been informed about the gar-
gantuan size and impact of the park.[7] Wind power would cause strife

and pitched battles between neighbors and within families across the isthmus. And in the communities surrounding the barra where the massive wind park would have been located, protests, blockades, vehicle heists, and raids by state police would carry on for years. Those who vocally spoke out against the works of the project would receive death threats and beatings. But with perseverance and strategies learned over decades of political unrest, protestors would ultimately stop the park's construction. It would be arrested and incapacitated. This would be a success.

The case of the Mareña park, in the scale of its potential and the enormity of its collapse, is an instance of one megaproject undone. But while the park has its singularity, it can also be taken as emblematic of programs of renewable energy that fail to deeply engage with and account for the people, things, and other beings that are coincident with them. The anticipatory good that the park was meant to bring, both for local development and for the energetic redemption of the global climate, existed as subjunctive futures: the might be, the could be. But that potential began to wither. It was not a series of technical flaws that presaged the giant wind park's denouement. Instead, its collapse was consecrated in the relationship between human hopes and an increasingly frail ecosystem. Wind power would have been a cleaner way to generate electricity, but the creators of the Mareña project failed to realize the ways in which their plan reproduced an extractive model in which "resources" are possessed and sold and the proceeds are divided, often in inequitable ways.

It has been the modus operandi of fossil-fueled modernity to extract resources in places that are relatively remote from the centers of consumption.[8] However, in mimicking the logics that have underwritten the carbon economy for the last three centuries, renewable energy transitions risk repeating old conventions that end in ruin. New ecoenergy forms ought to instead proceed with an ethos of rehabilitation rather than resource extraction. This should be an exercise in rebalancing human aspirations for energy with the energetic life needs of the more-than-human beings with whom we are in orbit. It ought to be a reckoning with forces like wind and water as well as an encounter with our technomaterial apparatuses. In truth, we cannot afford to get it wrong.

FIGURE INTRO.1. Wind turbines, Isthmus of Tehuantepec

Aeolian Arrivals

The power of electricity and green neoliberalism have converged in the isthmus, reshaping life in the region. Over sixteen months of fieldwork, our research team of two traveled to all the critical sites of wind power development in Oaxaca, from the isthmus where turbines were being sited to the country's capital where policy makers struggled to develop a program of energy transition that would be beneficial not only for the Mexican state but for the world's climate.[9] Our project became a practice of defamiliarizing "wind power" as a singular, technical, managed energy form, looking instead for the multiple ways that "aeolian politics" were gathering force. Aeolian politics—borrowing from the Spanish *energía eólica*, meaning electricity derived from wind power—emerged and evolved in many directions, from policy acts to the placement of bodies on barricades and from salvational winds to broken habitats. Aeolian conditions are everywhere in this work, expanding the term to mean many kinds of wind and their competing energies.

We set out to see what sorts of social impasses or collective victories were informing the terms of renewable energy futures. To do so meant understanding the positions of all involved, those commonly thought of as stakeholders. These were people living in the shadows of the turbines or on the threshold of a wind park yet to be born. And they were the land creatures and sea life inhabiting those same domains. They were renewable energy company executives and representatives of the Federal Electricity Commission (CFE), Mexico's national electricity provider and sole grid operator. And they were environmental professionals tasked with protecting watersheds and environmental systems. They were officials at every level of governance from local representatives in the isthmus region to state policy makers in Oaxaca City to lawmakers in the country's capital. They were journalists and laborers, aspiring politicians and hard-boiled *caciques* (local bosses), truck drivers and fisherfolk. And they were those who lived in and with the wind. We spent many hours with activists who were opposed to the parks as well as those who applauded the arrival of the *eolicos* (turbines). Over the course of our work, we also spoke with many, many "regular" people about their thoughts on the wind parks, on development in Mexico, on Pemex (the state-owned petroleum company) and renewable energy, on climate change and transnational capital. These were people we encountered in the course of our days, who might not have seen themselves as implicated in the political sweep of wind power or renewable energy but who were, nonetheless, part of a greater aeolian politics.

We went to where the wind is in order to grasp how renewable energy forms were coming to occupy the global South. But we also went to where the oil is. Mexico continues to be a petrostate—in recent times it has been dependent upon oil revenue for 43 percent of its federal operating budget. With declining oil reserves, however, the country had suffered financially, with much of its economic lifeblood buried deep under water in the Bay of Campeche.[10] Some regions of the Mexican state, however, are rich with wind, and in the early part of the twenty-first century, the country's policy regime tilted optimistically toward the development of renewable energy infrastructures. In fact, Mexico was among the first countries in the developing world to institute comprehensive climate change legislation, earning the country international accolades from environmentalists and industrialists alike.[11] If we were seeking to understand the phenomena of energy, Earth and human habit, we found that conjunction in Mexico: bioplanetary effects and the multiple energies that have fueled them.

Corporate investment and state sponsorship inaugurated the development of wind resources in the Isthmus of Tehuantepec beginning in the mid-1990s when the first wind park, sponsored by the CFE, became operational in La Venta in 1994. By 2004 a full-scale study of the entire wind corridor, devised by the United States Renewable Energy Laboratory, provided evidence of the considerable wind power potential that the isthmus held. Much of the region's land was marked "excellent" for the production of electricity.[12] During Felipe Calderón's presidency (2006–12), when the power of drug cartels soared and oil began to wane, a serious campaign began to develop the renewable energy sources of the isthmus. Although never compelled to do so through the Kyoto Protocol, new legislation in 2009 required that 35 percent of the nation's electricity come from noncarbon sources by 2024.[13] Lucrative incentives for private-public partnerships were created, and the Mexican wind power sector flourished. In 2008 there were only two parks, producing 84.9 megawatts of power in the isthmus. Four years later there were fifteen parks, producing more than 1,300 megawatts, making Mexico the second-biggest wind power producer in Latin America. By 2016 the wind energy infrastructure of the isthmus had expanded to twenty-nine parks with 2,360 megawatts of capacity, a 2,676 percent increase from the first years of operation. According to the Mexican Wind Energy Association, AMDEE,[14] Mexico's total installed wind power will reach 15,000 megawatts by 2020–22.[15] While these metrics are evidence of impressive and rapid growth in the wind corridor, they cannot begin to capture all the complexities of wind power. There is much more to it.

Staying with the Turbulence in Transitions

Wind power is not just any power.

It is a promissory force. Unlike mining, logging, or drilling for oil,[16] wind power generation is supposed to, in part, save the world. Infrastructural programs that claim to climatologically benefit the "greater good" hold a particular ethical ballast. Renewable energy projects would seem to righteously, and rightly, drown out the banal drone of greedy shareholders or demands for cheap fossil fuels. Wind power offers both redemption from dirty energy and, in places where wealth is sparse, the potential of economic salvation. But there is complexity all the way down. In many places in Latin American and elsewhere, denunciations against the environmentally destructive practices of fossil fuel extraction have now morphed into protests against projects marked by the ambiguous sign of "sustainability." Challenges have arisen as to whether local places are being sacrificed in the name of global climate salvation.[17] And yet resistance to anything that is environmentally "sustainable" or is a technology of "resilience" can be taken as suspect. From one vantage point, those opposing new-energy infrastructures can be accused of obstructing the future and gambling with unknown climatological consequences that are still evolving. By this logic, if local populations of people, and others, become irreversibly disrupted in the transition process, then that is simply the price to be paid. The global stakes are so high, and correcting the planet's faltering temperature equilibrium looms as the sine qua non of the subjunctive future. And yet old practices of extraction and exploitation can easily inhabit new spaces of sustainability, preserving a status quo that continues to seek cheap resources and vast tracts of exploitable land. Energy transitions thus beg the question, What precisely is being sustained? And what is being maintained?

Scientific consensus has determined that carbon incineration needs to end, but transitions have proven to be ambivalent. They are at once anticipatory and unknown. Hope is gathered here, but caution is too. Distinct scales of ecological remedy—those tuned to "local" worries or, on the other hand, to "global" concerns—can be incommensurate, each focused on addressing particular kinds of distress and distinct vectors of contamination. By emphasizing benefits to planetary ecological systems, local ecosystems may be further imperiled; and yet in failing to ameliorate widespread global impacts, the entirety of the living world remains in jeopardy. Therefore, in order to take an ethical position that prioritizes future possibility, it is important that we attend to how the mechanisms of transition are being operationalized, precisely because they can create their own forms of harm for humans and others. Each increment of ecological care ought to be thought of and enacted as a composition toward a whole. So while there should be no argument about the superiority of wind over oil in terms of externalizations and environmental damage, the institutionalization of any new energy form should inspire questions before resolutions.[18] And it has.

As I earlier wrote, this book was meant to narrate an antidote to the Anthropocene. And in some ways, it still can, but not without hedging and equivocating as to whether human beings can rebalance a warped world and restore habitability.[19] The Anthropocene speaks of vulnerabilities and risks, not simply for particular creatures, plants, and persons, but in the aggregate and in the future. A growing awareness can be sensed in dramatic weather events, such as cyclones and superstorms, just as it can be read across mediascapes in reports of fatal heat waves and arable land becoming desert. Anthropogenically induced environmental precarity will not be felt the same everywhere by everyone; the consequences will be uneven. Nonetheless, people around the world are increasingly exposed to the direct material and physical truths of ecological mutations and exaggerated weather forms. We are living it. And in this sense, the Anthropocene is not simply a geological designation for the human impact upon earth; it is a way of explicitly recognizing that impact. It is a state of consciousness.

As a planetary condition of precarity, the Anthropocene conjures a certain kind of extinctophilia, or an attunement to the necrotic. Each move that is made to chill the effects of a heating planet exists as an implicit recognition of human fragility.[20] Projects for sustainable energy that aim to mitigate further climate and biotic destruction are, in part, predicated upon the recognition that as a human species, we too are endangered. In ways like never before, "we" hang in the balance,[21] traveling the risky corridors of species being as the Anthropocene intensifies its effects. This means confronting extinction in new ways, that of other species and our own. But a state of impairment has a way of focusing attentions.[22] As Anna Tsing has described, life on a "damaged planet" is also a prerequisite for "livable collaborations," which are, in turn, the stuff of survival. If ruins are now our gardens and blasted landscapes compose the sites of our livelihoods, then we need to find optimism and perseverance in these ruins, in the cracks and fissures, in the spores and weeds.[23]

This is where wind comes in. Like the air out of which it is made, wind thrives on interplay with bodies, both lively and inert. An oscillation of gases and heat differentials, wind is an insistent reciprocal exchange between air, beings, and objects. Its relationality is important, even indexical. It is in contact that wind is seen. We might think of leaves quivering or branches undulating, dust in the air or a plastic bag aloft. In all cases, wind is seen only in those places where it touches or moves something else. A pencil drawing of curled lines is a way to illustrate wind, as are graphs, charts, and maps. But ultimately wind is only ever made visible through its impact and influence on other matter, other materials, and other things. Wind's ontology refuses to take separateness as an inherent feature of the world. Its relationality exists as an inverse allegory to the teleology of extraction that operates in one direction, to one end and for a singular purpose. And this is, in part, wind's value—it has an existential precondition that appears only in the context of contact. Wind is touching, mutual, moving.

In an era of renewable energy transitions, wind exists as a heuristic assemblage where powers and future imaginaries are tethered to one another. But wind also refuses to be gathered or to be caught as a thing; it cannot be held in a jar. Unlike other resources—such as water, land, or oil, wind evades enclosure; it is nothing if it is not movement, and therefore it is a force that is not easily made into a propertied object. Placed in a box, its ontic state is fundamentally transformed, becoming air. It is a force that may be captured but never contained.[24] While wind's kinetic force may be seized by the blades of turbines, wind in itself cannot be held. It is elementally loose. It is motion. Even as wind may be inanimate, it is nothing if it is not animated.

In ecologies of relationships that survive and sometimes thrive in the gusting winds of the isthmus, I want to avoid drawing deep divisions between the ontological capacities of nature and society and instead find their useful recompositions.[25] There is no fetishization of nature, or Nature, here. In fact we might begin with the acknowledgement that "nature" (or for that matter "environment" or "ecology") now exceeds and overspills definition.[26] Attempting semantic jurisdiction over the terms of what *is* or *is not* natural or constitutive of *the* environment is a conceit best left in historical place, like in mid-twentieth-century theories that lavished attention upon such binaries.[27] As Marilyn Strathern predicted a few decades ago, somewhat prophetically, our epistemic climate has increasingly come to represent an epoch "after nature."[28] These kinds of dissolutions and temporal demarcations seek new, re-adaptive thinking.

If there ever was one, the "thin bright line" between people and the mystified category of nature appears to be increasingly dissolving. Jackrabbits and Nissans, sand dunes and electric current, turtle eggs and stunted corn crops now all occupy this side of history, a cohabitational zone of socionatural space. Many thinkers have begun to emphasize the importance of recognizing the coconstitution of human and nonhuman beings, or what Donna Haraway calls "making kin." As the demarcations between humans and nonhumans have increasingly crumbled, in rubble too is the contention that "natural" history can be disentangled from the history of "Man." From this, theses have emerged prompting questions as to how "human" human history really is or ever was.[29] Where we have singularized human activity and separated it from everything else, we have, in fact, failed to understand the evolution of modernity and globalization as processes of interaction between material forms and forces as well as among multiple species. Of course, the history of capital must likewise enter into this genealogy because it has conditioned lifeworlds the world over. In this context, it has become clear that the "social" in social theory needs to be reproportionalized, at least the "social" that has been bracketed as referring to exclusively human interrelationships.

An Anthropology Alive to the Anthropocene

Some have suggested that the Anthropocene is a remarkable and unique gift to the discipline of anthropology itself.[30] Both nominally and epistemologically, Anthropology has claimed to be *the* science of anthropos, and its practitioners have spent well over a century attempting to grasp the many ways of human being. However, anthropological work has likewise been keenly alert to the conditioned specificity of "nature," particularly as a code that seems to surface most dramatically within its putative inverse: modernity. The bimodal categories of nature/culture and environment/society have sparked debate and challenged normative assumptions in the discipline for many decades. Such juxtapositions, their theoretical generativity, and the recognition of their limits have roots in philosophical propositions. But perhaps more importantly, they have been gained through empirical wisdom. The people with whom many anthropologists have worked, historically and in the present, often claim no rigid, exclusive, categorical distinction between human living and the material and multispecies domains in which people and their others interact and thrive.[31] From this accumulated insight, an anthropological fascination with a posthuman condition, multispecies studies, or more-than-human encounters would seem to be a rather "natural" outcome for a discipline that has observed firsthand the refusal of nature as a singular form.[32]

If the trouble with nature has been an anthropological preoccupation, displacing a universal understanding of *the* human might qualify as an anthropological obsession. Illustrating difference across human experience while also narrating transparticular similarities has remained at the core of anthropological work. Anthropologists have spent many decades demonstrating that there is more than one way to be human, and so it would seem the next step is to think through how the more-than-human is equally part of that story. In the conjunction of human and more-than-human encounters, attention to material things and other species should also encourage us to take humans as a species: a species that has altered earth systems and a species that faces its own status as newly endangered. Put another way, in a human-contorted world, we ought to push toward deepening the groove in which cross-species intimacies or socialites are evolving. Perhaps we are now even obligated to work beyond the human, as no element of human life exists untouched by ecosystemic circumstance. Where nature is increasingly erupting through human lives and vice versa, to ignore the unhuman is to walk willfully blind into a time of vivid possibilities.

The infusion of the more-than-human into the science of anthropos has not come sui generis. It has developed together with correlates in the physical sciences, from biology to physics.[33] Science and technology studies have stressed the incorporation of agentive technologies, machines, and apparatuses into human being, and this perspective has been woven into the analytics I use here. Feminist epistemologies, in particular, have provided generative ground for multispecies studies and techniques of science and technology. Attention to cyborgs, for example, explicitly called for the machinic to meet the biological, frustrating an easy separation between natural and "man-made." In times of ecological instability, the biological itself can likewise be recognized as more permeable, or "transcorporeal," as Stacy Alaimo has put it.[34] In the conditions of the present, I am especially cautious of delimiting our intellectual method to a facile version of actor-networked forms of agency.[35] In order to understand our environmentally precarious form of late industrialism, as Kim Fortun has reminded us, we must be responsive to the material and social ontologies of toxic conditions and unlivable environments that are not fully captured in actor networks.[36] Where discursive approaches to meaning have operated to distribute nodes of power and their outcomes, thinkers such as Karen Barad have also insisted that physical substance (matter) must be given its due in the world's becoming.[37] Or perhaps many worlds' becoming.[38] As she has put it, "Matter matters as much as mattering"; the physical and the significant are inseparable.[39]

The call to name this age the *Anthropo*-cene may appear to some as the ultimate aggrandizement of an overbearing species that is now carving its name into an epoch: the Age of Man.[40] However, as we well know, the Anthropocene condition did not come about through all people equally but from the cumulative acts of certain people with particular powers, the great majority of them being men. Past times that have valorized particular kinds of male achievement established a reigning Age of Men that, in turn, produced the Anthropocene age. And while the accumulation of human hubris may remain underfoot in plasticized and carbon forms in planetary stratigraphy, we can also aim to refuse the spirit of anthropos's reign. If an Age of Men created the Anthropocene condition, it is now time to invert that logic. Response to ecosystemic precarity will need to come from everyone, everywhere; it is not that fault lies equally, because the global North bears the greatest blame, nor that solutions will be evenly executed, for the global South is facing the greatest scales of harm. There is risk in flattening species being into one grand humanity because it erases histories of exploitation and futures of unequal consequences. However, debates about the qualities, origins and outcomes of the Anthropocene have invigorated questions about the place of the human in the world and the worlds that humans share with all other earthly life and things.

The Anthropocene may guide us to interrogate the consequences of a dominating anthropos; but, in truth, feminist thinking has always had that kind of attunement. The designation of an Anthropocene age invites a species reckoning to be sure, but it also summons gender trouble. This is a good thing. Old Cartesian distinctions that cleave human social and intellectual dispositions from their ecosystemic origins ought to continue to face critique. Equally important, however, is that the politics that have allowed for these sorts of inorganic fissures—which are almost always posed as natural—should likewise become part of a sedimented history of man that we leave behind.

That "anthropos"—as "Man"—resides so centrally within the notion of the Anthropocene is, in every way, an invitation to a feminist corrective; that corrective shapes the way I have written this book. Citational practice is one of the tools we have at hand as authors, and I use that prerogative intentionally here. While the scholarship in this book reflects a range of thinkers and disciplines, all of which are represented in the notes and bibliography, I prioritize feminist scholars of environment and ecological conditions in the text by using only their names in the body of the book. This is intended as a small counterbalance to the current politics of citation where male authors (often from the global North) continue to accrue more citational recognition, and thus legitimacy, particularly in the domain of theory. This is what Sara Ahmed has called "the citational relational." My intervention here is meant to acknowledge and surface a dynamic that unfortunately continues in the production of knowledge. This may be an imperfect experiment, but it is, from my point of view, a beginning.[41]

Fueling the Anthropocene

The Anthropocene speaks to the human manipulation of terrains, animals, and air. It calls attention to a process that has been ongoing and that may, in fact, singularize humans as a species. While people have always changed the land, creatures, and atmospheres where they have lived, we now live in times of exaggerated scale and depth. Humans grew up in the Holocene, an epoch that began almost twelve thousand years ago. It was in those conditions that we learned our agriculture and our letters, arriving at a state that we have been inclined to call civilization or culture.[42] But if the physical sciences have begun to agree upon the traceable existence of anthropogenic impact on earth systems, there remains disagreement as to *when* this Age of Man began and *what* it was that initiated it.[43] Propositions regarding the onset of the Anthropocene range widely. It may have begun with the age of agriculture or with imperial expansion to the (so-called) New World or with fossil fuels or with nuclear fission.[44] Each has its logical origins and outcomes. Holding these differences in place, what remains consistent about the Anthropocene are three things: it is about time; it is about exploitation; and it is about fuel.

 TIME

The suffix "-cene" is derived from the Greek *kainos,* meaning "new." But geological time is very slow and newness rare. And thus, the Anthropocene asks what it might mean to be *out of time*—chronically allochronic—incapable of imagining a seemingly boundless past, or an infinite future.[45] A new -cene might also attune us to what it might mean to be *out of time*—as in the jig is up and apocalypse is upon us. Distorted worlds may need troubled temporalities. And yet the Anthropocene continues in its accelerationist mode. Current extinctions are happening quickly.[46] We may worry about our own. These are times of prolepsis, where seeing a knife in the first act means knowing that the cut will certainly come. But we might also see a more hopeful foreshadowing here, where a grain of sand is a sign of a gem to come.

A fascination with Anthropocene causality and the periodizations of its unfolding is an indicator of one of the epoch's signature dynamics: bringing us deeper into our collective encounter with time.[47] The marriage of human history and geologic time is a call to the subjunctive form. We may recognize the future as both a lure and a tripping point because the Anthropocene is an anticipatory exercise. We know for certain that the skyscape is radically altered for millennia to come. *Geos* itself, with a seemingly infinite existence, embodies time that is deep and long. Temporal immanence like this can be cognitively challenging for those who live the fleeting existence of a human life-span. When Kathryn Yusoff writes that the Anthropocene is an opportunity to imagine ourselves "geologically,"[48] in the slow accretive metaphors of minerals and timescales in the hundreds of thousands or millions of years, she is correct. And yet in many ways we have already been living geologically. While the Anthropocene underscores how humans have become geologic agents the world over, our cohabitation with hydrocarbons and fossil fuels—harvested from deep down in geos—are an indication that we have long been, if unconsciously, already living geologically.[49]

Times that are marked under the sign of the Anthropocene may simply mean that the difference between life (bios) and nonlife (geos) is now more assertively marked, even as it has always existed and been apparent to some and not others. In Elizabeth Povinelli's reading, this historical juncture is no longer a matter of life and death. Instead the "new drama" being staged is a form of death "that begins and ends in Nonlife." Extinctions far and wide expose anthropos as simply another installment in a grander collective of not only animal life but all "Life" as opposed to the state of "Nonlife."[50] Traveling far enough back in time, we find that it was geos that supplied the conditions of bios's becoming. It was an "inert" earth that gave birth to all life. Just as humans have engraved themselves in and on geos, so too has geos permeated humanity in various ways: molecularly and biopolitically. The Anthropocene is therefore not only a way to locate the sedimentation of human practice, it is an invitation to uncover how bios has always been interlocked with geos.

EXPLOITATION

The colonial affliction that began in the middle of the last millennium forever altered the movements of people, animals, and plants. Worlds were brought together in unprecedented ways, often brutal but sometimes benign. At each step, residual marks remained on the crust of the earth itself. Exploiting lands and people at scales that were heretofore unseen, imperial conquests and settler colonialism induced bouts of growth and withering. The transformation of forests and farms into private, enclosed plantations was often powered by enslaved human beings and their forced labor. As wild places were replaced with plantation monocropping, biotic abundance and panspecies habitats became denuded: contorted places to grow plants for profit. What began in the colonial era as the radical transformation of diverse kinds of human-managed farms, pastures, and forests is in the present exacerbated by agribusiness and industrial meat production, in what Anna Tsing and her colleagues have dubbed the "Plantationocene."[51] These biotic shifts may have multiplied in the fifteenth century, but those effects were intensified further in the long sixteenth century and the rise of capitalism.

While capital may be famously promiscuous, humanity on the whole cannot be assigned equal responsibility for the injurious channels that it has produced. Anthropogenic harms that have accrued under the figure of "the Capitalocene" are a combination of capital accumulation and the (human) pursuit of power.[52] Operating in dialectical fashion, capitalism itself would become a world ecology enabled by changes in science, production, and the distribution of power. Capitalocene temporality, like that of the Plantationocene, resides in the extension of imperial seizures that engendered a restructuring of "natures" everywhere.[53] Plantations—and their close kindred in industrial agriculture—as well as capitalism continue to be at work in the here and now. As we recast history in the light of a deforming planet and climatological troubles, it is also vital to recognize that neither plantations nor capitalism nor industrial accelerations would have existed if it were not for anthropos. And that puts us back in the Anthropocene: a human-created epoch generating uncertain futures.

FUEL

The Anthropocene is often diagnosed as a plague of particular fuels and their burning. In the latter half of the eighteenth century, the European industrial revolution turned to new fuels and increased scales. Forests had been erased across much of Europe to sustain a growing population, but the machines of the industrial age required more efficient resources, and they were found in fossil fuels. The advent of the Anthropocene age can be traced, by some accounts, to one singular invention: James Watts's steam engine. This was the juggernaut of a peripatetic modernity, fed by coal that it would burn and burn and burn. Two hundred years later the industrial age reached its zenith and the mid-twentieth century would become marked as the age of more. Everything exponentialized: human populations, modifications to land masses, production and trade, excavations of petroleum pockets and mineral beds, the use of nuclear power for war and energy, and the emissions of gases and pollutants accompanying each increase in scale. They call these velocities of change the "Great Acceleration."[54] Speed and carbon formed an unprecedented coupling, and fuels became remainders residing in earth systems. Coal and oil, along with the split atomic nucleus, are the fuel forms that are most often associated with the Anthropocene and its accelerationist tendencies. But, I would argue, there is a critical other.

While petroleum certainly hastened anthropocenic conditions, we can also find the causal power of wind at work in the making of the Anthropocene. It was the power of wind that blew ships to the New World, inaugurating an age of imperial expansion and the increased exploitation of land and people, creatures and minerals. Wind-powered sailing ships transported goods back and forth, moving flora and fauna to disparate places, providing an aeolian infrastructure for the movement of people and things. It was

You can't possibly blame wind for the actions of humans → following its due course.

wind power and human greed that spurred the transatlantic slave trade. And through this set of abuses against anthropos came further intensifications of agriculture, continuations of displacement and the realignment of much of earth's matter. Wind that blew toward the New World led to certain kinds of futures. Captured within it, then and now, are other, potentially more equitable, possibilities.

Perhaps it is irrelevant to speculate on which time periods, social processes, or energetic sources can be charged with increasing ecological precarity. Does it matter, in the end, whether it was atoms or oil or wind at the root of it all? Maybe not. But maybe so. Unlike carbon fuels, wind has been positioned—by governments, industry, and environmental advocates alike—as a way to reverse the Anthropocene order. Therefore, while wind might be blamed for abetting the trouble on terra, it also embodies a response, a solution, or a method of energetic salvation. Wind thus holds us in an uncomfortable paradox: it exists as both partial cause and potential redemption for an anthropogenically wounded world.

This book is an attempt to live within that paradox by illustrating how wind fails when it is made to repeat the extractive logics that have sustained carbon modernity or, conversely, how it can succeed by giving its energetic potential not only as a source of power but as a source for imagining politics and ecologics anew.

THE FIRST CHAPTER OF this book is named for what it attempts to contain: "Wind." While elemental forces of air, water, earth, and sunlight have long maintained human and other life on the planet, they are now more broadly recognized as spheres that are at once crowded with extinctions as well as teeming with energetic potential. In this chapter, I engage with how wind is a dynamic and heterogeneous figure—a force of aeolian multiplicity—that is formed by land and by hope, by technocratic management and by human care. In this process, I argue, wind becomes differently, moving from element to condition and from experience to resource. Wind power itself can be said to occupy very different places in any map of the Anthropocene as a force that fueled the epoch as well as one intended to undo it. Wind's very ontology, therefore, calls for a "deterrestrializing" of thought, and what this chapter ultimately shows us is that wind cannot in fact be contained, only captured for a moment.

In "Wind Power, Anticipated," chapter 2, I track the evolution of wind power and its parks in the Isthmus of Tehuantepec. In this origin story is embedded the developmental aspirations of those who promoted the growth of wind power as a renewable energy source. In its most dramatic utterances, wind power was anticipated as a salvational object with far-reaching benefits. Out of this calculus came the Mareña Renovables wind park, which would have been the largest single-phase wind park ever installed in Latin America. I take the case of Mareña as paradigmatic of the challenges facing wind power in Mexico and, by extension, elsewhere. For those who promoted the project, its creation held enormous ethical potential not only to generate great quantities of renewable electricity but to provide social and economic development to the region. For those who stood in opposition to the project, firm ethical ground also upheld them: rejecting corporate megaprojects and the industrialization of their environment. What I demonstrate in this chapter is that origins matter to outcomes.

Trucks, the subject of chapter 3, would seem to be an unlikely nonhuman collaborator in the development of renewable energy. Trucks embody petromodernity in almost every way, from their masculinist stereotyping to their fossil-fueled metabolism. However, in this chapter I show how trucks are fundamental to the evolution of wind power: compelling the process, physically, politically, and often affectively. In empirical terms, they are always at work in the construction of wind parks or transporting the material goods for their operations. Trucks literally drive wind power: in the men they transport, in the politics they create, and in the hopes and terrors they foment. Trucks enable mutual communication between matter and form. As a temporal marker, trucks also occupy the apex of Anthropocene accelerations, and trucks therefore serve as "indicator machines" as well as "transitional objects"—expressions of human and machinic interplay that lie between petromodernity and a renewable future. This chapter makes the argument that technomaterial tools, objects, or artifacts, such as trucks, need to be taken as (a) consequential "matter" in understanding the ecosocial politics of energy transition.

In chapter 4, "Wind Power, Interrupted," I navigate the second part of the story of the wind park that never was. Although bolstered by powerful allies and drawing from all the forces of governmentality, developmentalism, and transnational capital, the Mareña project found itself irretrievably interrupted by accusations of trampling indigenous sovereignty and endangering other-than-human lifeworlds. For many supporters of the wind park, criticism of it was motivated by desires for personal financial gain. But for those opposed to the park, its collapse was a resounding victory against domination and displacement. Mediated across international news outlets and echoing through the channels of the Mexican nation-state, the Mareña project became a paradigmatic case, as one government official put it, "of how it should not be done." This chapter details those impasses to show that while the project may have intended to bring "transition" to the region—in the form of renewable energy and economic development—those protesting the park saw no such transition. Opposition to the project ultimately shaped a philosophical critique as to whether renewable energy is really anything "new" at all, especially when seen from the point of view of centuries of domination and militant responses to that domination. I argue that in the end, transition is nowhere an objective or neutral process but one predicated on subjective positioning.

Chapter 5, "Species," is an invitation to unthink species as a classificatory system of categorization and to instead *be with* species. In this chapter, human expressions of displacement—like fears about the loss of land and territory—find their analog in other species' displacements: from jackrabbits to sea turtles. Species life in the isthmus is qualified differently in the context of anthropocenic conditions and this is consequential to how humans diagnose, quantify, and seek to manage the species life that is wrapped up in wind. Humans are a powerful species within the figure of the wind: calculating measures of "environmental risk" in the offices of government agencies and making claims about which humans, animals, or plants should be allowed to thrive or die in the isthmus. The feminist philosopher Isabelle Stengers has called attention to the value scales associated with animal testing, and I similarly take species in the Anthropocene as a particular form of animal testing: trials for both human and nonhuman lives that currently hang in biotic balance.

In chapter 6, "Wind, in Suspension," Mareña's fate is sealed. Through the rise and demise of what would have been the largest wind park in Latin America, it becomes clear that the project suffered no technoscientific undoing but was instead sacrificed to the play of suspicions. Proponents of the park saw opposition to it as the work of troublemaking outsiders and political opportunists preying upon green capitalist enterprises, extracting bribes and mounting protests to enhance their own financial and political networks. For those opposing the park, its supporters were equally suspect: interested only in their profit margins, in the form of rents and contracts, and abetting the extraction of resources in a place keenly attuned to the privations of transnational capital. The giant wind park was conceived in the paradigm that its global climatological good would correspond with the ecological, economic, and social worlds that comprise human and other-than-human life across the isthmus. But as I show, failures of attunement ran deep: histories of insurrection and displacement were not given their due, and perhaps more important, the imagined futures of local residents were fatally ignored. While the wind park's destiny was tied to all manner of political maneuvers by caciques and corporate representatives, ultimately wind power would be drowned in the watery spaces between people and fish.

The conclusion to *Wind and Power in the Anthropocene* is a joint reflection on the collaborative research that is detailed in each of the volumes of the duograph, *Ecologics* and *Energopolitics*. In our final chapter, we look toward aeolian futures through the turbulent present of aeolian politics. In revisiting the years of research and analysis invested in this project, we return to the original premise that compelled us to the field, and to Oaxaca in the first instance: namely, the global necessity of adopting less catastrophic fuel sources in order to avert further anthropogenic harm and climatological insecurity. In revisiting this work, we also affirm more strongly than ever that renewable energy transition must be undertaken in a more fulsome way than it generally has been and that it must include the contingencies of both anthropolitical concerns and the more-than-human lives that energy infrastructures touch. Transition, we find, fails to achieve its potential when it is muted by the logics of extraction that have ruled the last several centuries. In the end, we do not merely need new energy sources to unmake the Anthropocene; we need to put those new energy sources toward creating politics and ecologics that do not repeat the expenditures, inequalities, and exclusions of the past.

↳ Hence, the importance
of understanding shared
history btwn develpmnt
↓
3
/ ...

1. Wind

The Afternoon's Finger

In some places, the dust never seems to settle. Wind finds its way every-where in Mexico's Isthmus of Tehuantepec, harassing the blue-black feathers of a wailing grackle, raising small stones from the road, insinuating itself against the blades of turbines to make electricity. Isthmus wind, like wind everywhere, is a negotiation between gases that are compelled across space and time by combinations of heat and cold differentials floating over land and sea, pressured shifts in directionality and potency. This is the physical-ity of the wind, its material life and its ontological being. Wind becomes contoured by objects in its path—mountains and hills, cliffs and stands of forest, buildings and creatures. It also willfully exercises its force upon these things: carving, cracking, pressuring, and leaving its ventifactual imprints. It draws our attention to points of contact and intraconnective incorporations; it absorbs contexts and conditions, and we often know it best through touch-ing (in) it.[1] Wind may be a relief from the heat, a force to struggle against, or a welcome bluster that blows smoke from our eyes.

The force of the wind has long been domesticated by human actors—through the milling of grain, flying of kites, blowing of ships across the seas. But industrial-scale electricity generation and the sprawl of wind parks are unprecedented, both in the isthmus and in the world. Wind is now being taken differently—not as it has been for millennia, but as a renewable "resource," or as "clean energy." As wind is increasingly cast as a valuable commodity,

FIGURE 1.1. An imprint of the wind, ink and paper

and as its powers are rapidly industrialized, so too does it undergo a reformulation of what it is. Newer evaluations and valuations of wind may not entirely eclipse the ways that wind has been known in the past, but there are, nonetheless, undeniable shifts in how wind is seen to work and for whom. "Wind power" is now designated as a force with the potential to redraft the energetic relationship between humanity and the environment; it has been made to assume a responsibility for global climatological care. Thus, while the wind may have always mattered, it has now come to matter in different ways.

For the ancient Greeks, Aeolis was the god of wind; across the isthmus, it is energía eólica—wind energy—that has come to occupy lands and sky.[2] By definition, aeolian imprints are those effects of wind upon geological and meteorological phenomena. But the winds that create ventifactual contours also shape people and places. In this chapter, I want to explore this aeolian multiplicity, showing how wind and its powers are formed by land, by desire, by technological management, and finally, by the care wind is afforded by some—indeed many—humans. This is a turbulent space. Wind is changed: from element to condition and from an experience into a resource that generates power and its effects. In the wake of wind, aeolian subjects are formed, and wind itself comes to be produced differently through energy aspirations. Aeolian life gets entangled with cosmologies and subjectivities, but it is equally implicated in ethical questions regarding sustainable development. Such refigurations between material, human, and nonhuman worlds require a crafting of political possibilities that move beyond material determinisms and social structural theories that have underwritten the industrializing logics of the past three centuries. Wind's very ontology calls for a "deterrestrializing" of thought.[3]

Pelting

In the town of La Ventosa, wind is a force that cannot be ignored. It comes in gusts and gales. It blows over eighteen-wheel semitruck trailers, and it causes some varieties of trees to only ever leaf and branch in one direction. It plasters clothing against skin, and it will have you momentarily lose your footing; its occasional calm is usually abbreviated. And it is for this reason, in part, that the town of La Ventosa is now completely surrounded by wind parks, in every direction and at the terminus of every street in this little hamlet.

For Don José, wind power has been a boon. Passing through the carport gate that separates his house from the street, he remarks on the quality of the wind at that moment. Knowing that we are not from the isthmus, he is no doubt certain that windward comments are a good way to begin a conversation. He offers that it is not bad today, just average, as he sets about arranging folding chairs on the concrete slab outside his front windows. Somewhere behind the wall is a young woman, maybe his daughter or daughter-in-law, who is preparing plastic cups full of *atole*, a sugary drink made with corn flour. Don José's home is relatively untroubled by the dust raised by the wind, a dust that saturates seemingly every place in La Ventosa. He lives on a recently paved street. The deed of *pavimentación* was carried out by the local government in collaboration with a wind energy company that has a park just on the border of town.[4]

Don José, a landowner who has leased parcels of his property to the wind power company, appears to be doing quite well. He has a large gate around his two-story home, fresh with paint. He attributes his relative prosperity to his contract with the company and to the monthly income generated from renting the land on which turbines and roads have been placed. Don José epitomizes the developmentalist dreams of wind power in the isthmus; his swelling wealth is imagined to flow in a trickle-down fashion to other, less fortunate residents—shopkeepers, laborers, and others without windy land.[5] Don José openly shares his story, situating it within a longer history of the town where he has always lived.[6] He is notably philosophical and methodical with his words, and his utterances are more ecological than most. After the atole has been drunk and we have been through our questions about the rise of renewable energy in La Ventosa, Don José turns us again to the wind. He wants us to know that the wind itself has made him strong. Like everyone in the isthmus, he explains,

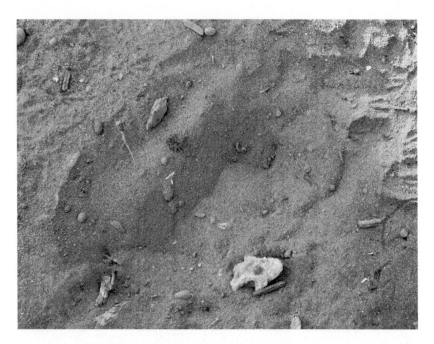

FIGURE 1.2. Road, La Ventosa

living with "el norte"—the powerful northern wind that whips across the isthmus from November to February—has an impact upon a person. "El norte picks up rocks, pebbles, and sand, and it hits you in the face. It gets everywhere. And you have to stand up against it and keep working and keep going in spite of it," he explains. "It makes you tough and unafraid." Don José is clear about the fact that the turbines on his land and the power of the wind have made him richer. But he also recognizes how the wind has formed him as an aeolian subject, a man who is abraded, contoured, and affectively shaped by wind.[7]

Air and Breath and Everything Alive

The north wind whips through,
in the streets papers and leaves
are chased with resentment.
Houses moan,
dogs curl into balls.

There is something in
the afternoon's finger,
a catfish spine,
a rusty nail.

Someone unthinkingly
smoked cigarettes in heaven,
left it overcast, listless.
Here, at ground level, no one could
take their shadow for a walk,
sheltered in their houses, people
are surprised to discover their misery.

Someone didn't show,
their host was insulted.
Today the world
agreed to open her thighs,
suddenly the village comprehends
that it is sometimes necessary to close their doors.

Who can tell me
why I meditate on this afternoon?
Why is it birthed in me
to knife the heart
of whoever uncovered the mouth
of the now whipping wind,
to jam corncobs in the nose
of the ghost that pants outside?

The trees roar with laughter,
they split their sides,
they celebrate
that you haven't arrived at your appointment.

Now bring me
the birds
that you find in the trees,
so I can tell them
if the devil's eyelashes are curled.

Víctor Terán, "The North Wind Whips"

Víctor Terán is a poet and a teacher.[8] I suspect he would put poet first when describing himself, but he is nevertheless a man who is interested in sharing his words and perspective on the world through both mediums: poetry and pedagogy. Víctor was not someone we heard about through the world of wind parks, but a man whose work had already been familiar to us because of his renown as a literary figure and a proponent of binnizá cultural and linguistic preservation. The place where we are able to meet with Víctor evokes neither of these qualities. Instead it is a bland, somewhat fussy restaurant in Juchitán called the Café Internacional. The café, so accurately named, has the somewhat dubious reputation of attracting Spaniards involved in the wind power industry as well as prosperous patrons from around the region. It is almost always a jangle of activity, with soccer games on televisions, waitstaff in prim uniforms, and a security guard patrolling the sidewalk. The Café Internacional is also one of the few places in a very hot town that can brag about air conditioning. This seems like an ironically apt climate for our talk with Víctor, which would ultimately speak of air, breath, and everything alive.

"You know, the wind has many meanings," Víctor begins. In Zapotec the word is *bi*.[9] And bi is what signifies the air and the breath. It is the soul of a person. And it animates everything. Linguistically, Víctor explains, the concept of "bi" is used to name all living beings. And it is for this reason that nearly all of the binnizá words used to designate an animal or a plant begin with the prefix "bi-."[10] Including *binnizá* (the people) itself. *Bini* represents a seed, its reproductive essence. And so it is possible to say that "bini" is the soul or the seed of a person, their inherent substance. Bi is an enlivening principle. It names the pig that makes the sound *bibi*, and it designates the worm, the maggot that crawls from dead flesh: *bicuti* is the creature that is both a product of spoiling meat and one that furthers decomposition of the flesh. "In this way," Víctor explains, "one can see that the Zapotec language is very metaphoric." But more importantly, he wants to emphasize, the concept of "bi" is inseparable from language itself; "bi" is etymologically inherent to expression in the same way that it is fundamental to life. "Without air, there is no life, and for this reason we use this prefix, bi-, for everything. It is very interesting, and it is very important," he continues, "because 'bi' is the soul, the air, the breath, and the wind as well. It is a bundle of meanings." Bi is more than a prefix; it is a repertoire of sensation and being.

Víctor depicts it plainly. "Without the air, we would not exist. Without the wind, we would not exist." The first animates, and the second is animated.

Cosmologically, there is a trinity of winds among binnizá people: two from the north and the other from the south. The first, *Biyooxho'*, the old north wind, *el viento viejo*, should not be mistaken for a feeble wind. It is, in fact, the opposite: the wind that made the world through its astounding force, its primal intensity. Biyooxho' is the northern wind with an ancient genealogy. At the beginning of time, Víctor explains, Biyooxho' "pushed the world into existence." A less storied wind, but one that all istmeños know equally well is *Biguiaa*: the northern wind that is quotidian and less dramatic but still insistent when it blows. And finally, there is the southern wind, *Binisá*, the wind of the sea and the water, a revitalizing and gentle wind that soothes the heat of the day. It gathers across the Laguna del Mar Muerto, just on the edge of the Gulf of Tehuantepec. Bi, air/life/breath/wind, is here married to *nisá* (the breeze) and in this union becomes moist. Binisá is often described as a feminine wind, a more tender sensation. Each of the northerly winds is inversely described as *masculino*. The gusting northern wind, Biyooxho', is also at times called "the devil's wind." Its heat and intensity make it seem as though it has come straight from Lucifer's lips. The winds of the isthmus accrue many powers of becoming and enacting, and it is wind and air that link body and cosmos, humans and deities.[11] "It is true," Víctor concludes, nodding, "there are many kinds of wind."

Wind is captured in a conversation and in cosmologies about how the wind makes people and what people make of the wind. Partly an oscillation of gases and partly an insistent reciprocal exchange between air and beings, the wind's relationality is essential. This kind of relationality, Karen Barad reminds us, produces entities as phenomena.[12] It is in these inseparabilities and intra-acting agencies that things and forces are configured as subjects or objects or relata. It is the wind's relationality that performs the work of creating aeolian subjects, who live in, from, and through the wind in its various formations and effects.[13] With attention to the ways that humans and our coinhabitants are drawn into wind and given life through its quieted form—air—we can pose the question, as Luce Irigaray has, as to whether "we can live anywhere else but in air?"[14] Like the air out of which it is made, wind thrives on interplay and incorporation, into and against, bodies. Captured by the meters of energy production but still residing in the domains of myth, legend, and experience, wind is wound into aeolian matters and their subjectivities.

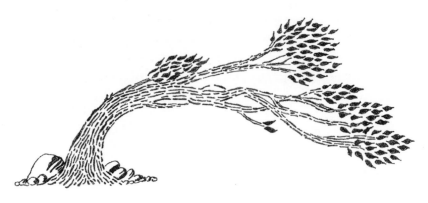

FIGURE 1.3. Tree and wind interacting, ink and paper

Industrial Densities

Whereas Víctor described the wind in terms of its sensations and its animating significance, the technical capture of wind has far more cartographic and quantitative explanations. In the early nineteenth century, a team of surveyors in the isthmus found "an almost incessant wind [that] either blew down or inclined obliquely the landmarks." It was a wind that caused their instruments to "oscillate violently" and disturbed their observations. And with this wind came a certain haunting. With the exception of a few moments before the rising of the sun and a few after its setting, the surveyors' chronicle continued, "a dense flickering vapour hid from view the objects which served as guides, whilst the refractions, especially the lateral ones, produced the most strange illusions."[15]

Far less enchanted than the nineteenth-century depictions, the 2003 report crafted by the US Department of Energy's National Renewable Energy Laboratory (NREL) sought to graph the quantitative details of how wind pushes its way through the isthmus. Barometric pressure differentials between the Gulf of Mexico and the Pacific Ocean are the essential source of wind in the isthmus. South of the Chivela Pass, passing through a fissure in the Sierra Madre, air from the Bay of Campeche flows from the north to the Gulf of Tehuantepec in the south. This is where wind blows its fiercest. Winter winds regularly acquire speeds up to fifty-five miles per hour, sometimes reaching tropical storm or hurricane force. Whereas Víctor associated the powerful northern wind with the origins of the world, the NREL report diagnoses this northerly flow in terms of pressure gradients. The Interamerican Development Bank (IDB), whose loans have been instrumental to the

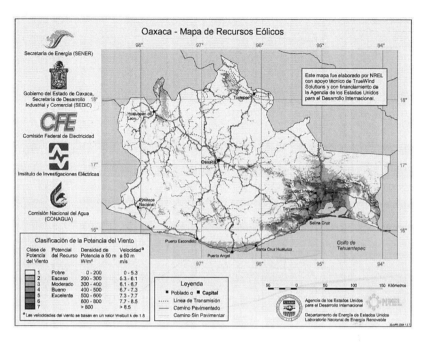

FIGURE 1.4. Wind resource map of the Isthmus of Tehuantepec

construction of many wind parks across the isthmus, finds the richness of isthmus wind using more terrestrial aesthetics, noting that the region is "a natural tunnel" for wind. Because of this, the bank can boast that the isthmus is "one of the best wind resources in the world," clearly "an ideal place for wind energy projects on a grand scale."

The North American company TrueWind Solutions and NREL utilized a computerized mapping system and GIS (Geographic Information Systems) software to track the wind of the isthmus. In conjunction with other entities—such as USAID (the United States Agency for International Development), the Mexican Secretary of Energy (SENER), the Federal Electricity Commission (CFE), and the Oaxacan State Government Secretary of Industrial and Commercial Development (SEDIC) among others—a vibrantly colored map showing the wind resources of the isthmus was brought to life.[16] Meteorological stations that tested wind quality were located on the Pacific Coast (in the port city of Salina Cruz) and inland (at Ixtepec). Station data was then assessed by wind power developers—such as the Spanish company Gamesa and the US-based wind energy company Clipper. The Federal Electricity Commission also weighed in on the information. With all expertise summarized, the report explicitly notes the proprietary nature of the data

System lacks transparency.

it exhibits. "Due to confidentiality agreements," it states, "we are not able to show the actual wind resource at the sites or provide the exact locations of the sites."[17] The derivation of the data and the precise qualities of wind in a given place and time have become questions of property, both present and future. Wind is re-formed—through numerical exposition and proprietary knowledge; it has become a commodity.

Seven wind power classifications color the wind maps of Oaxaca. Each of these categories is enunciated according to its "utility-scale application," ranging from "poor" to "excellent." Locations with an annual average wind resource greater than four hundred watts per square meter, or approximately seven meters per second—at an altitude of fifty meters above ground—are considered best for utility-scale uses. The measurement's height is important for the accurate accounting of wind speed, as are the effects of entities that might block, hinder, or tamper with its flow, such as trees, buildings, or towers. There are many material considerations, but NREL charts an optimistic cartography of the isthmus. Wind resource maps and other details contained in the NREL report allow companies and officials to identify prospective areas for wind-powered electricity to be generated. But assessments of the wind are also, inherently, a calculation about the land beneath it.

A little more than 7 percent of Oaxaca's total land area (91,500 square kilometers) is considered to be good, "windy land." The best wind resource areas are said to be concentrated in the southern reaches of the isthmus. High-quality winds for electricity generation bank from the southern coast, spanning sixty kilometers to the north and then another sixty to eighty kilometers from east to west: a cube of rich air. Surveyors have assessed that 6,600 square kilometers of Oaxacan land has "good" or "excellent" wind passing over it, with approximately two-thirds of that deemed to have "excellent" wind. According to some estimations, Oaxaca's windy land could support up to 33,000 megawatts of installed electrical capacity.

Knowing this about the isthmus wind, it is not surprising that Thomas Mueller, a German executive at a wind power company that has parks across the isthmus, was explicit in his response when we asked why companies such as Mueller's would stake so many billions of pesos on capturing the isthmus wind given the area's ever-fraught politics and the well-circulated narrative that the region is "ungovernable."[18] Given the territory's reputation for insurrections at every scale, from political upheavals to roadblocks, infrastructural investments such as those laid down by Mueller's company

FIGURE 1.5. Turbines and high-tension wires, La Ventosa

seemed, at best, a risky proposition. Although our question to him had to do with political conditions rather than ventifactual reasoning, his answer mapped immediately back to aeolian coordinates. "We are in the isthmus," he said plainly, "because it is the best resource in the country."

Reiterating what we had heard from other energy officials in the Mexican government, he explained, "Mexico has no subsidies for developing renewables in comparison to the US and Europe. So the development only becomes attractive when there is a *factor de plantas altas* [higher production quotient]. European levels of wind wouldn't be sufficient here." A report generated by Santander, the international investment bank with more financing in the isthmus than any other, also notes an important quality of the isthmus wind: its daytime quality. As opposed to northeastern sites in Mexico, where wind favors the night, the isthmus wind is diurnal. Therefore, the bank's accounting surmises, the available profit from electricity generation is higher in the isthmus than in the northeast because daytime usage rates are higher due to demand.[19] The attraction of the isthmus and the conditions that make its terrain and skyscapes sites of dense capital investment are due to the high production possibilities of its wind, in terms of both when and how much it blows.

In this estimation, the wind is a calculus, a quantifiable resource of a particular kind, that is harnessed for capital accumulation, growth, and profit. Wind can be prospected, and prospectuses can be made on and about it.

The qualities that have made the isthmus wind attractive for investors and developers of renewable energy have also made that wind an entity that demands protection, that demands defense, and that demands to be loved. The Assembly of Indigenous Peoples of the Isthmus in Defense of the Land and the Territory has protested vociferously against wind parks in the isthmus. These battles have been fiercely fought in both local actions and translocal responses.[20] How the assembly understands the wind and how they believe it has been appropriated are put simply and repeatedly in their statements and manifestos.

> ¡La tierra, el mar, el viento, no se venden se aman y se defienden! (The land, the sea, the wind, [they] are not for sale, [they are to be] loved and defended!).

On another flyer, on another day, amid months of protest:

> ¡Mareña Renovables, entiende nuestro viento no se vende!!! (Mareña Renovables, understand that our wind is not for sale!!!)

Echoing the calls of environmental protection that are familiar in the global North and elsewhere, the assembly has vowed to safeguard the wind, to affectively embrace its vulnerability as well as its powers. As Maria Puig de la Bellacasa reminds us, caring is an ethically and politically charged practice; it calls upon an affective state and a mode of engagement that comes to form an ethico-political obligation.[21] Here, the wind of the isthmus is a frail object that requires human care and attention, protection and love.

Inasmuch as care becomes a form of practice to enable a relationship with the wind, in equal measure the assembly indicates their commitment to decommodify the wind and the land beneath it. For the assembly, wind is not a resource that can be sold, and they have explicitly disavowed the notion that wind can be captured under the regime of private property. Against the prevailing neoliberal philosophy that underwrites much of Mexico's development apparatus, including wind power, the assembly refuses a monetization of the wind. In so doing, they likewise refuse to take the wind as a transactional resource. Yet in caring for "our" wind, there is also the sense that it is an entity that can be secured. Despite its resistance to enclosure, wind is here held and claimed, as "ours."[22] Wind, then, may repudiate captivity in capital terms, but it does not eschew human stewardship. Wind may not be a commodity or even a resource, but it is nonetheless discursively animated in the domain of human interests. For the impassioned critics of the wind parks, the isthmus wind fully belongs to an "us." It is "ours."

Aeolian Possessions

Facing an audience of several hundred seated in the collective meeting hall in Ixtepec, Sergio appears primed to give his report.[23] Having left his native Spain, Sergio has been advocating for years that the *comunidad agraria* in Ixtepec be allowed to build a community-owned wind park.[24] Already several hours into the meeting on a Sunday morning, it is Sergio's task to rouse sentiments, and ultimately action, despite the swelter of the day and the room's overcrowded and poorly ventilated interior. The aging comuneros— communal landholders of approximately 114 square miles of Ixtepecan property—have gathered for another round of debates about the wind: how to create a community wind park on comuna lands. Sergio's invective hurtles over a sea of cowboy hats that are being used as fans for weathered and perspiring faces. The electricity commission, CFE, Sergio announces, continues to block the farmers' plans to construct a community-owned wind park on their collectively managed land. Worse still, the comuneros' most recent attempt to break down the bureaucratic barriers that prevented them from building their wind park had not worked. But the fight is still worth waging. "The issue," Sergio shouts, "is that the rights that are being violated here are the rights that are being violated in communities across Mexico. It is not only in Ixtepec; it is in all these communities. Because CFE, basically, what they are saying is, 'You don't have a place here, even though it is your land, even though it is your wind, and even though it is your (electric) substation . . .'" Taking a breath, he continues, "You don't have a place. All of this is for the multinational companies."[25] Shifting in their seats, the comuneros appear taken with Sergio's words. They too know that their wind needs to be defended against foreign capital brought by the *gachupines* (Spanish) and gringo (North American) interlopers. "Our wind" was shaping into something that could be stolen because it was becoming something that could be possessed. —> once it became a commodity!

To whom, or to what, does the wind belong? For the governor of the state of Oaxaca, the wind, or more accurately, *el aire*, was "public property"—a resource to be distributed across Oaxaca's citizenry through the mechanisms of development, job creation, and electricity. For many of the residents of La Ventosa, wind was held with more uncertainty. When we polled the inhabitants of this little town, many of them expressed their belief that the wind appeared to belong to the wind power companies, at least more so than to local residents. It was an energy supply that was being extracted. It was not a public good but a privatized one. Wind had exceeded the grasp of the people, its material figure instead claimed by the rotors of company turbines. But wind among La Ventosans was also never fully tamed. For most of them, wind belonged to God. Or to no one.[26]

Vitalities

The blades of a turbine both capture and displace wind, seizing its kinetic energy while also shifting its direction in centrifugal patterns back toward the sky or the earth. As the turbines lumber through their patterned acrobatics, there is a sonic dimension to the interplay between the wind and blades. A massive respiration and then stilled quiet. Picking our way through the brush and weedy patches between the turbine towers on a January day, we finally come upon what our host wanted us to see. Fernando Mimiaga Sosa is, at the time of our meeting, the director of sustainable development for the state of Oaxaca, and he, perhaps more than anyone else, has influenced the trajectory of wind park development in the isthmus. He is quite well versed, very well connected, and definitely a character. Fernando is someone with whom we will have many conversations.[27] But for now, during our first full day together, Fernando has asked his driver to pull over at one of the wind parks outside the town of La Venta. He called ahead to one of his friends to be sure that the security guard would allow his little Toyota sedan through. We are up close to the turbines now. They are immense and imposing and impressive, and they have a warning sign on them: DANGER.

Fernando, however, wants to show us another kind of danger, and that is the impact of wind upon crops, corn in particular. After some glancing around, he finds the sort of specimen he is looking for, a tiny exemplar of a corn plant, with the characteristic leaves, but maybe half the size it ought to be. The stalk itself is what he is particularly eager to demonstrate. Bending down, he shows us with a flattened hand how stunted the stalk is, short and ungrown. A failed plant. "This is what the wind does," he assures us. "It won't allow these plants to grow to their full height, to mature. It blows them over. And corn is supposed to be tall," he adds, gesturing upward. "And this one, you see, is *chiquito, muy muy chiquito* [small, very, very small]." It is true. While corn production has decreased in importance for isthmus agriculture,

FIGURE 1.6. Turbine tower with DANGER sign

ÁREA ENERGIZADA

PELIGRO

that fact is, at least for what Fernando intends to communicate, beside the point. For him, it is the wind that has diminished the corn plant, its force and velocity have hindered its growth.

And there is something special about corn. It is emblematic of the Mexican national imaginary, from precolonial times to the present. Corn is present, ingested, and (literally) incorporated into the national body everywhere in Mexico essentially every day. So when Fernando shows us the pathetic state of corn here in La Venta, he is assuredly gesturing to something else: a deprivation and a lack of economic growth in the region that ought to be remedied. And this, we will see, is part of his mission in the industry of wind-generated electricity. The wind, which has stunted the corn, can now be redeemed; it can be made to grow the economic potential of the isthmus in the form of wind parks. Later, back in the Toyota, Fernando concedes that wind development projects in the isthmus "began without adequate legislation and without the grid and necessary networks in place." "But," he emphasizes, "the most *hermoso* [finest/loveliest] thing is the wind itself." He explains, "The wind in the region continues to go on. Excellent. If we compared this wind to a gem, we would have to call the wind here a 'diamond.' It is the wind that is guaranteeing development; if it weren't for the wind, there would be no development."

The wind, for Fernando and certainly others, contains the riches necessary to transform the isthmus from a place of limited economic horizons to a place of developmental triumph. He recognizes that boons like these, like wind power, must be managed in order to sow the collective good. But for many others in the isthmus, apprehensions surround the turbines that have come to populate their lands and territories. And while corn may be the metaphor and the material entity that Fernando is using to make his case for the benefits brought by wind parks, another comestible concern is beginning to emerge in the shadow of the turbines.

In January 2014, *Las Noticias*, a major daily newspaper in Oaxaca, featured a story about "machines that eat the air." The journalist interviewed two brothers, Emilio and Juan, who counseled the reporter on several changes that they believed had been caused by the wind parks. "Before," Emilio described, "the nortes [northern winds] of the Dead Sea zone were tremendous." He went on,

I don't know, maybe they were 150 kilometers per hour or so, knocking down trees and houses. . . . But now that they have installed the wind parks in Salina Cruz they are much less. And the shrimp and fish catch has been reduced. They say that now there is much more sorghum, corn, and watermelon because the wind no longer whips the plants, but the fishing has gone down; its production has reduced for the same reason, because there is no wind.

Emilio elaborated this condition further.

When the norte is here, it jostles the seawater [making it turbid] so that the shrimp, fish, and crabs don't know whether it is day or night, and so they spend the whole day floating [near the surface]. Then we can go out during the day and we catch them, and we go out at night and we catch them. But now, now that the water is clear, they don't come out. And before, the norte would last a month, and now it is only a few days and it is gone. So we believe that what we had before here with the wind, now it is being absorbed by the turbines.

When the norte is at full force, the water itself is changed. And for fishermen such as Juan and Emilio, the helpful powers of the wind have been altered and diminished by the turbines. Has the norte been absorbed into the blades of the *aerogeneradores*? In a metaphoric sense, it has; the powers of the norte have been diverted. Where it used to serve fisherfolk's wants and needs, the wind now gives itself elsewhere, to the industrial production of electricity for corporate consumption. Where the wind once allowed these fishermen to contravene the usual state of creatural awareness and thus easily pluck shrimp, fish, and crab from the sea, now that has become less fruitful. Another excavation has been put in its place: reordering the wind to serve electrical desires rather than the watery wishes of isthmus fishermen.

Whether or not the turbines in fact "eat the air" is one question, but the fact that they are believed to is indicative of other questions and concerns. Turbines occupy a semiotic field, and their towers function as a location for anxieties and apprehensions. Emilio and Juan might have claimed a different origin for the changes they observed in the water's color and clarity. They might have just as easily pointed to climate change writ large: how it is altering weather patterns, including the direction and duration of the wind. But they do not. And it is not likely for lack of knowledge about the fact of a changing climate, for this is a fairly common discourse in the isthmus. The reality of *el calentamiento climático* (climate warming) and stories of how *el clima ha cambiado* (the weather/climate has changed) are the subject of conversations among those who work the land and sea in the isthmus.[28] Instead of man-made climatological warming, the changes in the wind seem to be even more directly produced by humans: by machines, which are powered by capital, that appear to eat the air. Unlike the abstractions of climate science and rising seas, the blades of wind turbines make present certain kinds of materiality. They are there for all to behold and to serve as a reservoir of climatological, maritime, and biotic unease. The "white giants," as they are often called, form an apparatus that establishes this climactic reality in ways that the rather ineffable designations of "climate change" or "global warming" cannot.

Passing through the central square of Ixtepec one afternoon, a billowing children's bouncy castle on our left, a girly-magazine stand on our right, our friend Raul Mena wondered aloud. Had the winds in the isthmus changed? He was concerned, he said, about how the turbines and the rapidly reproducing tracts of parks might alter the ways in which pollination would take place. For those plant species that required seed dispersal and pollination on the back of the wind, how would they survive or how might they perish? And, of course, there was the question of birds, which were also responsible for

FIGURE 1.7. Sorghum seeds

spreading seeds in numerous directions and flows. How might their routes change, or how might they be harmed over time and irreparably?[29]

Luis Zárate is a man who has become quite famous for his art. He builds his aesthetic works from the environment he inhabits, depicting in organic forms the intermarriage of water, soil, and biota. He, like Raul, was concerned about the wind parks in Oaxaca. He understood well their purpose as climatological remedy, but he was uneasy about them. Disputes about land and money he understood as a given. But what was more mysterious and poignantly troubling for him was how plant species' reproductivity would be transformed. Seeds carried through waves of wind—"anemochory," the most basic form of seed dispersal—might cease to be, he worried aloud.

How wind is changed by turbines and how wind has determined the fate of plants, animals, and humans across the isthmus, in the form of stunted corn or lost fish, reveal elements of the wind's powers. These are concerns about the wind, in and of itself, but are nevertheless linked to human and other-than-human intrarelations. Wind is an entity that does something for humans—allowing for a generous catch. It serves nonhuman species as well, propagating plants and allowing for their flourishing or not.[30] And in this

sense, the wind signals a coeval resonance among humans and their others, an elemental medium of relationship.

Ending (in) the Wind

Air in motion may be barely felt, or it may overturn and mangle homes, bodies, and machines. From some cosmopolitical viewpoints, wind is life and spirit. Víctor Terán knows bi (air/breath/wind) as life force, feeling, respiration, which is fundamental to all life, animal, human, and otherwise. Experiencing the wind as a comingling with bi holds a sublime appeal: it makes wind/air/life whole, integrative, and mutually evolved. But in the isthmus, wind is also often known as a force unto itself. It is the *animatum* in the case of bi, or it lashes rocks against skin and hardy spirits. It forms aeolian subjects, enriching them or training them to endure future pains. It is routed through lives. For agents of government and industry technicians, wind is a quantification: a commodity and a metric achieved through the grand schemata of science and its calibrated gauges. Wind can be divided and distributed in more or less equitable ways. For some people, wind can belong to the companies; it can be owned. And for others, it belongs to God or to no one. It is patrimony; it is to be protected, given rights, cared for. It has volition, a material conative will: the ability to change people's fate, fetter the growth of plants, or conversely, have them flourish.

A kinetic commons such as wind is motile and dynamic, unsettling and unsettled. But wind is also that which touches us. Wind exists as a relationship among humans who negotiate value, access, and outcome. Perhaps more importantly, wind creates a relational domain between living and non-living beings, and it embodies a refusal of separation.[31] Wind is kinesis and air as interactive forces—between heat and pressure, to be sure, but also in relation to the world (or worlds) it touches. It is known through its contact and its fleeting connection. These kinds of relationalities, which produce a deeper sense of inseparability, may allow for a different way of knowing material life. Wind is this kind of matter. Through the blades of turbines or felt across the skin of those who inhabit the places where it blows, wind has a place. Potentially a very significant place in redrafting the possibilities of ecohuman futures in the Anthropocene.

In this chapter I have drawn out the ways that wind is multiple. In the coming chapters that multiplicity will be elaborated through the ways in which wind power, as an energetic force, becomes a nexus around which land and livelihoods, politics and patrimony, are gathered. Turning our attention to the huge wind park proposed by Mareña Renovables, we are also turned toward how the multiplicity of wind becomes directed, routed, and channeled for particular purposes with certain environmental and economic outcomes. As wind becomes captured in the apparatus of renewable energy, it likewise becomes swept up in grander environmental discourses as well as local struggles for recognition. And as wind becomes domesticated in the revolutions of turbine blades and is moved and shaped by human intervention, so too does it push and transform the very places and people with which it comes into contact.

2. Wind Power, Anticipated

This is the story of a wind park that never was.

Had it been constructed, the Mareña Renovables project would have been the largest wind park in all Latin America.[1] It would have prevented the emission of 879,000 tons of greenhouse gases, and it would have represented a new scale and dimensionality to wind power in the Isthmus of Tehuantepec, providing renewable power equivalent to the usage of more than 600,000 Mexican households. The proposal for the Mareña park had all the signature elements of success in a time of neoliberal development projects and attempts to remediate climatological harm. It was bankrolled by international development banks and touted as a categorical boon for local communities by government representatives—from officials in Mexico City and the state capital to provincial authorities in the isthmus. All its environmental credentials appeared indisputable, from infinitesimal attentions to ecosystem protection to the colossal demands of climate remediation. It was a project that would be the vehicle for massive quantities of sustainably sourced electricity, and the developers had attained dozens of permissions and reports, from environmental impact assessments to archeological preservation certifications to easements for roads, transmission lines, and subterranean cables. However, the park was not to be.

When viewed through the lens of furthering renewable energy transitions and fostering greener forms of power, the story of Mareña Renovables is a tale of epic failure. However, when seen through the eyes of those who protested the park's development and ultimately prevented its construction, the story of Mareña Renovables is a tale of sweeping success. In this chapter, I describe the origins and evolution of wind power in the isthmus, locating its developmental objectives—and sometimes machinations—of those who promoted its growth. I offer a sketch of the conditions of Mareña's emergence as well as the specific objectives the project hoped to achieve. The goals of the Mareña park were ambitious, perhaps audacious, not only because of the magnitude of the installation itself but also in how it traversed several communities, political factions, land tenure systems, and ethnic alliances and rivalries. Most perilous of all, however, was the way that Mareña Renovables bound its future, and ultimately its collapse, to the monetization of the wind as well as the land beneath it and to payments— or payouts—that were intended to facilitate its transition into being. In the self-portraiture of the Mareña project, we also see the seeds of its demise, its aborted existence. This is a story of a paradigmatic paradox: a transition that failed to transform itself from an extractive ethos and thus failed to realize the potential of new political forms that might come with new energetic forms.

For those who promoted the park, its construction held incredible potential not only to generate large quantities of renewable electricity but to develop the region economically and socially. Equally true, however, is that those who opposed the park also believed themselves to stand on firm ethical ground in their rejection of corporate megaprojects and the industrialization of their environment.[2] Each set of actors could claim moral propriety.[3] The state, the companies, and the opposition all believed themselves to be following the primary mandate of the ethical actor: to do good.[4] And yet each constellation of interests was both suspicious of and suspected by the others as having ulterior motives. These worries about duplicity and deceit began to rot away at any moral nucleus that might have been, sometimes quickly and sometimes slowly, but assuredly. Thus, while the Mareña Renovables project was effectively left in ruins through a combination of factionalism, greed, violence, and hubris, I will suggest that the denouement of the giant wind park was driven by the play of suspicions: each party's lack of faith in the ethical truth of the other. The wind park was a moral battle played out in a field of treachery. In this chapter, I chart the fractious ethical positions and begin to think through their consequences, both in the case of the Isthmus of Tehuantepec and in greater global ecological and political dimensions.

The Road to the Wind

When the phone rings at four in the morning, it can be a troubling sign. The bracing call in *la madrugada* (that time before dawn) in July 2009 was not trouble, but it was a rite of passage, a trip down the rabbit hole, and the start of the longest day of fieldwork ever. We had met Fernando Mimiaga Sosa and his son for breakfast the day before, and although we had read several of his presentations online, all detailing policy and engineering considerations regarding wind development in the isthmus, this had been the first of what would be many, many meetings with Fernando.[5] The breakfast briefing had typified many first encounters in fieldwork—slightly awkward, especially as Fernando's adult son attempted his jagged English while the father tried to maintain the conversation in Spanish. The interview had gone well. Nonetheless, it was surprising to hear from Fernando at four o'clock the next morning. Perhaps we had been too quick to celebrate our first important interview, or maybe we had just been foolishly indulgent in exploring the artisanal wizardry (and inebriating effects) of the region's famous cactus liquor, mescal. Whatever the case, we had stumbled back to our hotel only a few hours before the alarming ring. "We're leaving right now, Ximena," the voice said.[6] "I am sending my driver to your hotel to pick you up. You have to get ready and packed. He will be there in ten minutes. This is Fernando." The sounds were there, but language seemed particularly distant at this hour and in this state. The message was clear, however, and we were off. Fernando Mimiaga Sosa is not the sort of man who takes no for an answer.

What we had no way of knowing then, as we poured ourselves into the back seat of Fernando's little blue Toyota, was that our sixteen-hour journey would provide a template, a map, and a microcosm of all the windy politics in the isthmus that would occupy our time for the next four years. From bleary Oaxaca City stoplights to the flat plains of mescal cacti just outside the city limits, we rise into the Sierra Madre, our path lit by moonlight and flickering headlights. Fernando is beginning to shower us with details and facts that he might have missed the day before. He has a bottle of beer in hand, and it is not clear whether he had slept the night before. Luckily, he is not behind the wheel on the narrow, serpentine road that winds its way to the isthmus. There are many stops on our expedition. Each punctuated the cartography that make up Fernando's bimonthly trek to the isthmus: the place where he was born and raised and where he now was, in many ways, the king of the wind. A couple of hours into our drive, we stop at an open-air roadside eatery and feast on a saucy plate of deep-fried pork skin crowned with fried eggs. Then come two more stops for unknown reasons. Gas. A delay here and another there, a purchase at a convenience store, some phone calls. We have a more extended sojourn with Fernando's mescal producer. This is a one-man-and-one-ox operation, with the harnessed animal lumbering forward, pushing a heavy wooden pole that drives a mill that, in turn, grinds scorched cactus spines into potent liquor.

We come out on the other side, onto the flatlands, rolling into a little town where a dam used to be, and where Fernando's *mujer* (woman) lives. She gives us a cold drink, and Fernando asks whether we agree that she looks a little like an American starlet. Over the miles we have traveled, Fernando has explained his role in the creation of wind energy in the isthmus. Fernando, like other boosters of wind power, pointed to its many advantages. For one, the turbines occupy relatively little terrestrial space, thus allowing the land to be used for other purposes such as grazing cattle or growing crops.[7] Indeed, Fernando put some emphasis on how installing turbines and access roads would not reduce the ability of cattle ranchers to continue raising cows or cause farmers to give up their crops. If anything, Fernando saw the overall improvement of roads, and extensions of those roads, as an infrastructural boon that followed from wind investment in the region as companies sought to enhance the material lives of populations in and around the wind parks. Wind power could thus be neatly layered over, but would not displace, the activities of farmers and ranchers who had made their livings from the land for many centuries.

For communal landholders or private-property owners, three types of payments could be expected if turbines and roads were installed, Fernando explained: a payment for land use, a payment for any impacts on the land, and a payment calibrated to the quantity of electricity produced and then sold. Remuneration would occur in three separate phases of development: at the point of contract, at the point of development, and at the point of electricity generation and sales. In addition to these financial incentives, a whole raft of other potential benefits would be on offer from companies building parks in the isthmus: scholarships, community centers, health service brigades, and the sponsorship of soccer teams. In the case of bienes comunales, payments would, theoretically, be distributed to the entire comuna through the auspices of the *comisariado* (land commissioners); this would also be the case for ejidos still operating under collective management. In the case of ejidos that had been privatized, payments would go to the now-private landowners. There were goods to be had, either individually or collectively, depending on which company one was in business with.

In his time as the director of sustainable energy from 2000 to 2010, Fernando navigated the laws and policies such as they were. He also charted new ones.[8] Fernando explained how his office had organized international conferences with wind developers and experts and how he had worked with federal entities in the mid-2000s to expand the grid and the number of electric substations in the region. His idea to create a super substation in Ixtepec, he said, was what led the Comisión Reguladora de Energía (CRE, the federal agency that oversees the national energy sector) to institute the necessary measures, collaborating with private developers and CFE to make it so.[9] Through his wind work, Fernando ensured his own well-being, politically and likely financially—although that remains speculation. But Fernando also hoped, he said, to promote well-being and progress in his native land. Although he had for years occupied posts in Oaxaca City as a favorite son of Mexico's revolutionary party, the PRI, he saw his role in wind as a kind of homecoming. Indeed, loyalists to the PRI (or PRI-istas, as they are often called) had many accolades for Fernando. Members of opposing parties were not always so generous in their praise. He was a politically polarizing figure, but no one questioned his influence in shaping the way the wind would go in the isthmus.

One act for which Fernando was credited (or condemned), for instance, was the regularization of land deeds (*catastros*) across the isthmus.[10] Where there had been vagaries regarding property lines and historical questions of rightful possession in the early days of wind development, Fernando and other corporate interests had seen it as their task to designate property ownership, linking names and signatures to pieces of paper and parcels of land. The logic of private development demanded this. Companies wanting to sign agreements with landowners insisted on assurances that they were contracting with the actual owners.

There are, however, many contingencies as to what constitutes "ownership." The Isthmus of Tehuantepec, like other regions in Oaxaca, and Mexico more broadly, maintains two relatively robust communal property regimes, bienes comunales and bienes ejidales, which date back to the Mexican Revolution. Both land systems demand collective decision making in all matters regarding changes to the disposition of land.[11] Article 27 of the 1917 Mexican Constitution ensures usufruct land rights as well as a significant degree of local control over land use, which is to be decided by the membership (the asamblea). Bienes comunales, which were established to preserve (or reinstate) indigenous landholdings, and bienes ejidales, which provided mestizo peasants with land to farm, were in some cases able to recover traditionally indigenous land from large colonial estates. In Oaxaca the number of bienes comunales and bienes ejidales is particularly high, with 823 ejidos holding 18 percent of the state's total land and 716 comunidades/comunas holding 67 percent of the state's land.[12] Some of the best land for wind development in Oaxaca is maintained as communal property, and while some ejidos have elected to adopt neoliberal land reforms that allow individual landholders to make private contracts with wind companies, others have resisted. As ejidos were parceled in some places and communal land management was strengthened in others,[13] the ratcheting up of rural governmentality that wind parks brought would result in dramatic contestations about land, wind, and sovereignty.[14]

In his own estimation, Fernando functioned as an intermediary to facilitate the benefits of economic development and renewable energy in the isthmus. Although he had a residence in Mexico City, he told us, he worked hard on the ground, day in and day out, in a windy corner of the country in order to bring about change. He guaranteed that energy regulators, the Federal Electricity Commission, investors, and development corporations would all be satisfied. This involved, he explained, "negotiating with everyone on the ground. From the landowners to the company representatives to the state and the federal officials."

The art of Fernando's negotiations became clear soon enough when we stopped the car on a rutted road in the little town of La Ventosa. Even in the short walk from the car to the porch of the evangelical church where a meeting was being held, the heat was debilitating. Sitting in tiny chairs on the veranda were a couple dozen women, wearing traditional Zapotec clothing, fanning themselves with papers they had been handed; men stood nearby, keeping to the shade and waiting for the discussion to begin. There was concern about the payments that local landowners would be given for the turbines on their land. Representatives from Iberdrola, the Spanish renewable energy corporation that ran the park, were there to try to assure landowners they were being fairly paid. This was one of many processes of negotiation, and it was riddled with apparent frustrations on all sides; it was also a harbinger of the way things went in the isthmus.

The company men, pleased to see Fernando, were beaming and patting him heartily on the back. "This guy," they said, "he has been working with us for ten years now. This guy is half gringo and half *Chilango*."[15] Although Fernando did not appear bothered by being designated "half gringo," this was exactly the sort of cozy alliance with Northern powers that had roused ongoing suspicions of neocolonialism among residents in the region. Fernando

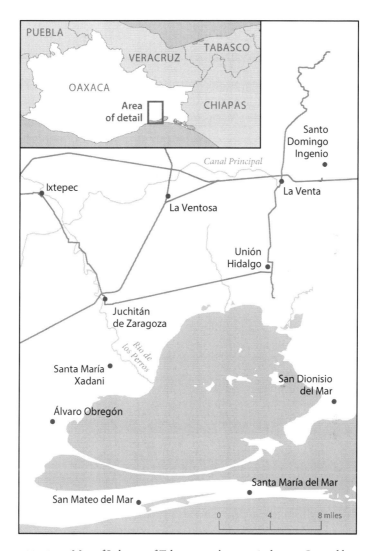

MAP 2.1. Map of Isthmus of Tehuantepec's core wind zone. Created by Jean Aroom with assistance from Jackson Stiles and Hannah Krusleski.

knew he was regularly and publicly accused. "You will see these things on the internet, put up there very quickly and in an alarmist way, saying that I am surrendering or handing over the country to the transnationals; but this is simply not true." The Spanish executives were certainly pleased with Fernando's work. Several landowners with whom we spoke later would also praise Fernando's ability to get the region signed on to wind power. Fernando kneaded environmental and social conditions into a shape that would be favorable for renewable energy and a new form of development in the isthmus. His ethical purpose—and his disposition, he believed—was to be a diplomat. He saw himself as a mediator. But perhaps, in less flattering terms, he was a fixer.

Infrastructure of Aspiration

Some isthmus winds blow so fiercely that they cannot be tamed by turbines. For the most part, however, istmeño wind is the steady pulse that keeps aeolian dreams alive. The electric side of the equation, that is, the apparatuses of infrastructure and planning, are less consistent.[16]

In Mexico, there is only one way to obtain a regular flow of electricity, and that is through the grid of the Federal Electricity Commission. A parastatal corporation that holds a monopoly over the country's current, CFE is tasked with supplying electricity to the entire nation, from lower-income residents (whose bills are subsidized) to commercial customers (who pay relatively high rates for their power).[17] But there are many weak links in the electric chain. A lack of power lines, especially in the central and northern regions of the country, leads to bottlenecks that prevent the proper evacuation of wind energy into the grid, according to the Mexican Wind Energy Association (AMDEE).[18] Wind parks in the isthmus have been especially vulnerable to CFE's lack of transmissional capacity. Although the commission is given few resources to do so, CFE has as its mandate the creation and maintenance of the grid, that critical vehicle through which Mexico's electricity is transported. And this is one very important reason that the logics of private renewable energy production have predominated. Although it was CFE that built the first test park in the isthmus, the institution has not invested robustly in developing the region's wind parks. It has, instead, elected to allow the sector to be privatized, with corporations bearing the costs of construction and installation.

According to the director of the Comisión Reguladora de Energía, there are two distinct drivers of renewable energy in Mexico: the high (commercial) cost of electricity and the country's exceptional solar, wind, and hydroelectric resources. The electricity commission must buy the least expensive power available and provide electricity to the Mexican nation at a fair rate. When the federal government considers the construction of a new power plant, a public tender is called by CFE, and the winner is determined based on cost per megawatt hour. What this means, effectively, is that any renewable energy project has to compete against conventional energy sources within a particular price context; this can be a difficult proposition when the global price of oil is low. In order to encourage private investors to develop power plants using renewable sources, the Comisión Reguladora de Energía needed to create different formulas in lieu of participating in the general tenders.[19]

Space in the substations was thus cordoned off for wind, and this, in turn, established a means wherein private-sector developers and CFE could enter into a temporary public-private partnership for the sole purpose of developing a new high-capacity transmission infrastructure. The system facilitated the conveyance of electricity from wind parks that had been developed by the private sector to flow into the national electricity grid. Although peppered with nebulous policies and expectations, the system, such as it was, functioned rather like a kludge, an inelegant assemblage of provisions that nonetheless functioned, for a time, to get wind power and electric infrastructures in place. As one investment bank document put it, "Whilst by no means perfect, with portions of the regulations vague or missing, it is a major advance over nothing."[20] Moreover, it followed a neoliberal ethos to the letter: the government was disbursing its public responsibility to private capitalists, who were, in turn, chagrined to do the infrastructural work of the state and yet, at the same time, were compelled by a desire to profit from the wind.[21]

This legislative, technical, and energic structure was why the secretary of energy and Fernando Mimiaga Sosa each promoted and instituted a model of self-supply energy production for the wind resources of the isthmus: *autoabastecimiento*.[22] As detailed in the 1992 Public Electricity Service Law, the autoabastecimiento model requires that the power producer and the power consumer are co-owners of the project. According to corporate financial calculations, the "only way to turn a Project activity or other renewable energy alternative into a feasible proposition is to create a Self-consumption

Company."[23] Corporate self-supply, autoabastecimiento, also means that the companies that have agreed to purchase the electricity (the partial co-owners of the plant) are able to buy that power at lower-than-market rates, usually for a period of twenty years. Any excess energy that is not used by corporate consumers—such as Walmart, CEMEX, Coca-Cola, FEMSA, Heineken, and the gigantic baked-goods manufacturer Bimbo—can be banked in a "virtual storage" scheme that is managed by CFE. Real-time electricity production in this structure does not have to match real-time consumption. Electricity generated but left unused by the corporate consumer/producer, has to be sold back to CFE at a fixed price.

The infrastructural advantage of autoabastecimiento is that the Federal Electricity Commission is able to auction off space in substations and often can oblige companies like those constructing wind parks to augment or build the required technical extensions and mechanisms that carry the electrons from place to place. Or, put more bluntly, as one Mexico City journalist specializing in the energy sector told us, "The [companies] feel like they are getting a shitty deal from CFE. CFE makes them pay for their own transmission towers and for the substation. . . . They aren't making much on these projects but then again where else are you going to find this kind of wind?"

 Comprising about 75 percent of wind power development in the region, autoabastecimiento has come to be the default model for isthmus wind. As a form of energic management and financing, it has assured at least three outcomes. It has fomented the privatization of wind-powered electricity production in Mexico. It has all but ensured that the renewable electricity produced will be consumed solely by corporate partners rather than local residents or municipalities. And, finally, it has compelled private developers and investors to augment a teetering infrastructure that the state has not been willing or able to subsidize.

A Piece of Cake

Traveling from place to place, negotiating with representatives and companies interested in developing the wind potential of the region, Fernando—perhaps not single-handedly, but certainly crucially—was able to facilitate a cartel-like arrangement of corporate interests across the isthmus. A map, which he included in his many presentations to investors, landowners, and bureaucrats, shows the region broken into districts, outlined and marked by company nomenclature. In these cartographic regimes, communities that lie

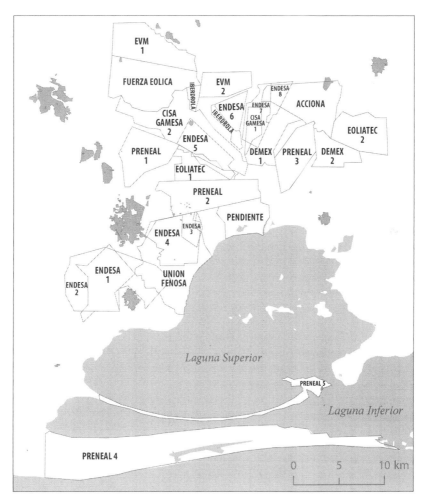

MAP 2.2. Corporate map of isthmus. Created by Jean Aroom with
assistance from Jackson Stiles and Hannah Krusleski.

within the boundaries of each corporate domain are only able to negotiate with that company, thus inhibiting an individual's or a community's ability to negotiate rental rates or contracts with competing offers. The map became infamous when it was appropriated by the antiwind resistance as proof of neocolonial extraction in pictorial form. It served as convincing evidence that the spoils of the isthmus were being cut up and apportioned like so many "slices of cake" (*las rebanadas de pastel*), as many put it.[24]

Government personnel like Fernando were not the only agents reshaping property regimes and producing documentation that would better facilitate the growth of the wind industry in the isthmus. International financial incentives, such as the United Nations' Clean Development Mechanism (CDM), were also critical to the development of isthmus wind. One application for access to the CDM program specified that standard loan financing for a wind park would be nearly 11.5 percent. The document went on to report that without the sale of CER (certified emissions reductions, also known as carbon credits) provided by the CDM program, projects would be financially unfeasible.[25] The application made the case that "no wind farm in Mexico has been developed without the CER's incentive," articulating that international financial support was crucial for wind development. "It is clear that the wind farm could not be developed without the incentive of the CDM registration due to technical and economical [*sic*] obstacles," one report noted.[26] However, the Mareña project, along with others in the isthmus, had also become more financially feasible in late 2009 when the Comisión Reguladora de Energía instituted a "postage stamp" (or universal willing) model of transmission rather than the per-kilometer charge of previous times; this allowed for a onetime, less costly route to the grid.[27]

Renewable energy companies themselves were also gathering signatures and contracts for land use, usufruct rights, and agreements regarding the use of roads and facilities. One enterprise in particular, the Spanish company Preneal, was very successful in assembling signatures from residents and expanding its reach across the region. Preneal was able to acquire some of the most lucrative land in the isthmus, including the site that would be slated for the development of the Mareña Renovables park. While Preneal had a small holding in the isthmus, the company never managed to actually build a park there.[28] In fact, where Preneal's talents seemed to lie was in the acquisition of land rights in the most profitable locations. Rather than engineering renewable energy sites, they were technicians of speculation, shrewdly acquiring signatures, permits, and authorizations.

After acquiring access to the sites that were to become the proposed location for the Mareña project in 2004, Preneal seems to have done little to maintain contact with the communities with whom it had contracted. Payments appear to have continued apace, at an annual rate of 250,000 pesos (approximately $19,000), paid to the municipal authority in San Dionisio del Mar to be invested in community development. However, Preneal—then operating under the name Vientos del Istmo SA de CV—neglected to maintain an adequate number of people on the ground to be in close contact with the residents who would be affected. When Preneal sold the development rights to the Mareña group in March 2011 for $89 million, these failures in communication became an increasingly irreparable impediment to the project. According to the financial bulletin documenting Preneal's sale, the purchasers were a "Mexican business consortium" (Fomento Económico Mexicano, or FEMSA) and the Infrastructure Fund of Macquarie Mexico.[29] In February 2012, after securing $700 million in new debt financing, the Macquarie Mexican Infrastructure Fund adopted two new majority owners, the Dutch pension fund PGGM and the Japanese corporation Mitsubishi. With this, Mareña was made. Following the autoabastecimiento model, the FEMSA group and Heineken-owned brewery Cervecería Cuauhtémoc Moctezuma were co-owners and would be the sole recipients of the electricity that the park would produce over the course of their twenty-year self-supply offtake contract.

After obtaining a tidy sum for flipping land contracts, Preneal all but disappeared from the isthmus.[30] The departure of Preneal, in addition to changing corporate ownership and management, exacerbated feelings of skepticism regarding foreign capital. Isthmus residents had seen many companies come and go, and wind developers seemed to have a habit of revising company names, further sowing distrust among local people. Much corporate nomenclature was afloat in the isthmus when it came to wind power development, and it was often unclear what the distinction was between the names of parks themselves, the owners of those parks, the companies purchasing electricity, and the sources of loans or financing behind the projects. As the new owners of the nearly $90 million package—which would ultimately swell to a proposition worth more than $1 billion—Mareña Renovables would inherit these misgivings and failures of transparency. The plot they bought was, from a terrestrial and windswept angle, a gold mine. But from a narrative and metaphorical point of view, the plot they bought was a tragedy.

There were consequences to flipping companies and land. Between the days of Preneal's speculative contracting and Mareña's final acquisition, fallen lines of information between the corporations and residents took a toll on the potential project. Although Preneal had allowed a communication gap to form, Mareña's management continued it, with that deficit leaving too much unsaid and unknown among local populations in the isthmus. No one knew the consequences of this communication breakdown better than Jonathan Davis Arzac, the CEO in charge of the Mareña project. There was also no one who could better profile the ambitions of Mareña than the man who effectively signed the checks. But with that power came the responsibilities of maintaining the ethical profile of the company and answering for its faults. At the Foro Internacional de Energía Renovables (FIER, the international renewable energy forum) in fall 2012, Davis would attempt to do both.[31]

Davis would make his speech in an utterly magnificent colonial setting: inside the walled courtyards of the Templo de Santo Domingo, a sixteenth-century church and former monastery that is visible from nearly everywhere in Oaxaca City. This is home to the botanical gardens where giant cacti and clutches of indigenous plants serve to augment the ecological portraiture of the second annual meeting of FIER. The governor, Gabino Cué, inaugurates the event with a series of renewable energy targets that the country should aim to achieve. He explains that this means pursuing a global direction and a goal. As the demand for electricity rises in developing countries like Mexico, Cué says, the sources of clean energy, such as those in Oaxaca, "represent a driver for the global economy, one that would produce more jobs while also reducing pollution in our seas and skies." Noting that in the past year the winds of the Isthmus of Tehuantepec accounted for 98 percent of the country's total wind energy output, the governor expresses that "we are obligated, . . . given our clean energy production, to create spaces for dialogue and reflection where each of the involved parties can add to the conversation and set the agenda, scope, and goals that we will need to optimize in order to incorporate the needs of communities and their families."

The FIER conference in late September is well attended by hundreds of people, many of them students pursuing degrees in renewable energy at the Universidad Tecnológica de los Valles Centrales de Oaxaca. In the front rows are the polished Mexico City bureaucrats and executives, often flanked by striking young women. It makes for telegenic news. And, as the title contends, it is international too, with renewable energy experts from Belgium, Brazil, China, Colombia, Germany, the Philippines, Singapore, and the United States. On the dais for this exceptionally well-attended panel is Jonathan Davis Arzac. He is here to clarify the purpose of his project and to promote its moral and economic worth. Given that everyone in the audience is aware that the project is in deep trouble, he is surely also here to defend it.

In Davis's performance are signs of nearly every challenge that burdens the park's future. By this time, criticism against the Mareña project is becoming well known across the isthmus, and voices of protest are gaining in volume. National and international media are generating commentaries, interviews, and diagnostics on the park's future. Davis's speech should be understood, then, in one dimension, as an effort to forestall further critique and to respond to the frailties of the project. But more than anything, his remarks are an attempt to weigh the park's moral legitimacy over and against complaint and contestation. These are his words:

> Good morning, everyone, and thank you to the government of the state of Oaxaca for the invitation to participate in the events today. I think it is worth it to take some time to talk with you all a little bit about the project that we are working on, what we're trying to do, and what sort of impact it will have. The company is called Mareña Renovables, and we are trying to construct what will be, once it is installed, the largest wind park in Latin America. It has a capacity of 396 megawatts, and the shareholder consortium that has come together to finance the project is formed by three different groups, each of which has committed about one-third of the capital investment. The first group is the Mexican infrastructure fund Macquarie, which is a [pension] fund that is dedicated to Mexican investments in infrastructural projects. It works only in pesos and to responsibly invest all workers' savings for the workers who have contributed to this fund.[32] Until now, the largest investment that this fund has made is, in fact, this wind park. The second shareholder is a consortium that is a Dutch business—who we know by its letters, PGGM, and I tell you only the acronym because it's practically impossible to pronounce the name of this fund—which has very lucrative holdings in Holland. And the third group is Japanese. You know them by the name of Mitsubishi, which produces everything from pens to elevators to cars. We are a very highly diversified consortium. The total amount of investments that we will make is one billion dollars, and to do this we depend on

financing that has been authorized by a group of banks, including foreign banks—approximately eight of those—in order to complement the capital investments that we are making.

We have obtained all the required permissions and licenses, and there are essentially two categories of these permissions and licenses. The first have to do with environmental issues; these are permissions granted by SEMARNAT [the Mexican Secretariat of the Environment and Natural Resources], and these are certified only after presenting documents regarding environmental impacts. There are also applications that are required for "land-use change" in a forested region, and there is also the issue of access and right-of-way. The negotiation on all of these projects is fairly complicated because one must negotiate on an individual level with each of the landowners. And the truth is that they continue being landowners, with the added ability to lease their right-of-way or access to their land, and so they can have this. In our case, we have already conducted about 270 negotiations that have resulted in the very basic ability to have a right-of-way for the transmission lines that will carry the energy produced in the park to the substation owned by the Federal Electricity Commission, where it will be incorporated into the flow of energy, into the grid, where the Federal Electricity Commission keeps an inventory and oversight over the country's electricity. I want to say that we have received great support from our governmental institutions, starting with the federal government, up to the presidency of the Republic. Whenever there have been doubts or troubles along the way, we have gotten incredible support, indispensable support, from those I have mentioned. For this we are also very thankful to the government of Oaxaca and particularly the governor, Gabino Cué. We have gotten support from local authorities like the municipal president of Juchitán [Daniel Gurrión], who has come to be with us here today. We have also received support from local authorities, such as comisariados [communal land commissioners], and without the help of these authorities, these organs of government, well really, none of these projects would be possible.

I want to also add and to explain to you a bit more about the project to give you an idea of how much will be saved in terms of energy and in terms of contamination. Almost one million tons of carbon dioxide each year. One million tons of carbon dioxide. We consider this to be a very noble project because it is not a project where some benefit at the expense of others. The companies that will consume the energy that is generated by the park will be able to receive a lower price [on their electricity] than the Federal Electricity Commission would normally charge, and thus they are going to have savings. And in addition to this, because they are taking part in the development of clean energy and renewable energy, [the companies consuming the electricity] are able to demonstrate that they are socially responsible companies, and so those who invest in these companies are also winning. As the professor said just a minute ago, the investors will also receive a very attractive return in order to compensate them for their investment risks, especially in consideration of the magnitude of financial resources that are being put toward the project.

Also, I want to talk about the communities, and here, perhaps, is where most of the myths have emerged in the past. The communities win, not because of the employment opportunities that these projects can generate, although it is, of course, important to look at the direct employment opportunities as well as indirect employment opportunities. Nor is it simply a question of the rents that are paid by the company for the right to use land. The park is also beneficial because of how we have modeled our project: we have transformed communities into partners. That is to say that there is another party that is going to benefit from the resource dividends the park will produce in the sale of energy, and that is the community. And the community will have the power to decide for themselves what to do with those dividends. We were just listening right now to the municipal president and the local congressman about a complaint that continues, and that is, How is it possible to have these wind parks that generate tons of energy, and yet we can't pay the electricity bills that are sent to us by the Federal Electricity Commission? And what I would like to say to you—that you have to see—and that is that you have to pinpoint this problem. Everyone in this country who uses some service or another has to pay for it, and here we need to ask if the prices are fair or not. But this is an issue for the Federal Electricity Commission itself and the secretary of energy, to say what to do with these profits and whether part of that should go to pay some of the [electricity] consumption costs in the communities. That's where the conversation needs to begin.

We have as our philosophy to respect the *usos y costumbres* [traditions and customs] of the population of Oaxaca and of every other country in the world, and we have established the highest principles, the best principles at the international level. We are not only committed to the Equator Principles and the World Bank and the guidelines that are provided by the Interamerican Development Bank. I will tell you here that we use these as the very basic starting point; these are the minimum requirements for us. We are a company that understands that success for a business like ours truly depends on the population feeling comfortable [with us] and that we are respectful of the environment. That is, it would be suicide to try to do something other than this. We wouldn't last even two years in operation! The relationship that we are looking for with the state, with the communities, and with the people is a long-term relationship where all of us benefit. We think that it is important to respect the environment and the fishing grounds of Oaxaca, from which populations have lived for many, many years and continue to live from. We are not here to see these areas affected. We are respecting the lagoons, respecting the vegetation of the mangroves. What we need to have is the community by our side, and we can, all of us, live together in peace and in a harmonious manner. We can achieve a common good and make an important leap forward in terms of development in these communities.

I admit that there have been many times, in our case, that we haven't been able to let people know what it is that we were doing and what we want to do. And this [communication] is something that we need to think of as a continuous project. Right now, we are relaunching our entire communication strategy, and I want to tell you that we are always open, every day of the week, in order to respond to any questions. We are a very open business, a transparent business, and a business with integrity. We have nothing to hide, and we're happy to answer any questions about the challenges in front of us. And it's important for me to return to the communities that are participating in the project so that the communities feel as though the project is theirs, and that this is your project, and that it is going to benefit everyone, and that it will allow your families to be able to attain better levels of well-being. Thank you.

Where Davis begins, with the prevention of massive amounts of greenhouse gas emissions, is critically important for him to convey. Where he ends, with the sentiment that communities must be incorporated into the development process as partners—not simply as workers or landowners—provides a crucial counterbalance to the ethical portrait of the park. Davis's narration of the project speaks to the many moral underpinnings he believes it to have. He is very aware of his audience, to be sure, nodding to his hosts and the governor and providing a plethora of obligatory citations to those institutions and governmental entities that have allowed the Mareña project to reach the levels of approval that it has had up until this point.[33] From the point of view of the state, as he makes very clear, the project has been authorized. All its papers are in order, and Davis is careful to acknowledge the ways in which bureaucrats and other state agents have facilitated this process, making them into nominal partners of the project as well. Davis also appears keenly aware of his larger audience, for throughout his discussion, he indexes the value that the Mareña project will bring to the people of the isthmus, the people of Mexico, and the people of the world. Social and financial benefits are interspersed everywhere in his speech, accounting for everyone: investors and officials, landowners and future families.[34]

Davis impresses upon his audience the financial magnitude and beneficence of the Mareña project. And the numbers are staggering. Staggering in the sense that the project represents an immense amount of investment on a scale that is unprecedented in the region. Although mining and hydroelectric dam projects have been carried out in the area,[35] requiring huge sums of capital, this is to be the largest financial investment ever made for a renewable energy project in the isthmus.

The fiscal risk is also staggering. In monetary terms, the Mareña deal has been constructed as a financial product, a tranche—risk is distributed across several parties through structured financing. It is also, at this time, the largest tranche ever made for a Mexican wind park. Emphasizing how financial resources are being tilled into the soils of the isthmus is clearly important for Davis and others advancing the project. They believe, or at least they say they believe, that this project will improve the quality of life in the isthmus and position future generations to aspire to new levels of growth and accomplishment. Davis is certainly someone who is professionally committed to the logics of financial growth. Before taking the position as the director of the Macquarie group, he was president of the Mexican central bank. He is a financier with fiscal prowess. In his speech, he briefly mentions—and possibly only because the issue is raised by another speaker on the panel—that investors themselves can expect a lucrative profit from the Mareña project, or what he calls "a very attractive return." Although Davis does not reveal any exact percentages in his presentation, we will learn in our conversations with banking officials who are knowledgeable about the financial terrain of isthmus wind projects that the rate of return for investors would be approximately 12 to 15 percent, a remarkably high margin of return on investment.

perk \ will benefit the
rich more than locals
through generating profit
for them

Companies purchasing the renewable electricity generated by the Mareña park, Davis underscores, are proving themselves to be conscientious actors. Investors in the Mareña project could likewise claim social responsibility, for as he points out, it is specifically a "Mexican" pension fund (that is being managed by the Macquarie group). This fund, consisting of the pooled resources of Mexican workers, would also serve as ethical leverage. Mexican workers' money would be financing an infrastructural project that would not only give pensioners a high rate of return but would benefit the region where the park was to be built. And, of course, the world at large would benefit from lower carbon emissions. But there is also ambivalence about Mexican labor here. On the one hand, Davis highlights that Mareña is the largest investment that the pension fund has ever made. At the same time, Davis emphasizes that isthmus residents will not be exploited for their construction labor nor will they be dispossessed of their land. Rather, they will be true "partners" in the process. While the ideals of partnership are a key element of Davis's professed goals, the challenges of working with multiple communities with different land tenure systems and ethnic alliances will ultimately prove to be a challenging model of partnering.[36]

Alliances are a critical trope in Davis's narrative. Invoking collective governance and financial backing provided by multiple institutions serves as a reminder of the project's soundness. Its adherence to the Equator Principles (a management assessment framework used by financial institutions to account for and manage environmental and social risks) serves as further proof of good standing. The project's international imprint is clear down to the corporations (like Mitsubishi) and financial institutions (a Dutch pension fund) that are financing the work. Even as Davis jokes about the "unpronounceable name" of the Dutch fund, this is no doubt in order to distance himself, at least to a degree, from the appearance of having too cozy a pact with the Europeans. Davis knows his audience. It is the Mexican nation and its governing entities.

And it was the audience, or at least some vocal representatives in the audience, who finally called attention to all the ways that the Mareña project had failed to live up to the aspirations that Davis had detailed. At the close of his speech, hands shot up across the crowd in a fervent effort to take the stage. The organizers of the FIER conference were then loudly and repeatedly accused of failing to invite representatives from the isthmus out of fear that they would be critiqued (which they were). Davis faced a series of pointed questions about the company's failure to acknowledge the will of indigenous people and local populations. Complaints rang out about how local residents had to pay a high price for electricity in the isthmus while giant cement companies obtained cheap electricity from the parks. "What real benefits have we seen," one woman asked, "after all these parks have been erected?" The environmental impact reports and other studies, one person claimed, were all in the service of the companies. Or, as another woman put it as she rattled off a series of doubts about Davis's portrayal, "Something smells bad in the isthmus."

Davis stood his ground. It was clear that he was a man who was, or had become, accustomed to facing condemnation. But as soon as he was freed from his place at the dais, he was off, quickly. So quickly, in fact, that we had to literally run to catch him. He and his colleague did stop and pass us their business cards. Davis's colleague, as the card showed, was Sergio Garza, author of Preneal's CDM application a handful of years before (as Fernando Mimiaga Sosa's son would wryly comment later, "Like Walt Disney said, it's a small, small world"). Davis told us he welcomed a visit from us at their office in Mexico City and restated the company's willingness to answer any and all questions. Over the following months, however, we would call many times and send several messages to fix a meeting time in Mexico City, all to no avail. Like the Mareña park, that meeting seemed doomed never to be. We would see Davis two more times over the ensuing months: once as he hurried to a semiclandestine meeting with a state official and again as he was ducking a crowd of angry istmeños in the lobby of a lavish skyscraper in Mexico City.[37]

Propaganda

The Spanish word for promotional materials of any kind is "*propaganda*." This lexicological lacuna allows for very easy double entendre. It was therefore important whenever anyone asked whether we had seen the Mareña Renovables promotional materials, to listen very carefully to the intonation that was given to the word "propaganda."

The promotional PowerPoint presentation that Mareña produced was a series of slides noting the project's benefits near and far. However, there was one particularly striking image, an almost-surreal computer-generated image of what the wind park across the sandbar would look like were it ever to come to fruition. The sands of the barra are almost alabaster in this image, unlike the grayish hue of the actual grains. And across the crest and arc of the sandbar are positioned, at regular intervals, gleaming white turbines that disappear into the distance. An aeolian vanishing point. The waters of the lagoons that ripple up onto the shore in the picture are crystalline rather than the turbid waters of actuality. It is beautiful, almost too beautiful, as it draws attention to the ethereal qualities of the place and the would-be industrialization of it. For the purpose of promoting the redemptive qualities of the park, the image verges on counterintuitive, the turbines spoiling the beach that stretches out to the horizon line. But the PowerPoint document is, nonetheless, informative in terms of the rote presentation of facts on each of its slides.

According to the presentation, the project will comprise 132 turbines, each with a capacity of three megawatts, meaning that the total installed capacity will be 396 megawatts of gross power. Of those, 102 turbines will lie across the Barra de Santa Teresa, with the remaining thirty situated in the town of Santa María del Mar. To evacuate the electricity generated, the project will require that 1.5 kilometers of cables be placed under the lagoon floor, which, the presentation points out, will not interfere with fishing activities. The transmission lines that will convey the electricity to the substation in Ixtepec total fifty-four kilometers of cable, wending their way across the region and through the village of Álvaro Obregón. Although the permissions for the project allow for the construction of six docks alongside the inner perimeter of the sandbar, the company, apparently in an attempt to lessen its environmental footprint, promises it will only install two docks, which are required to service the traffic of barges delivering the turbine mechanisms to the sandbar. Transport of materials to the barra during the construction phase has been carefully calculated to "minimize the effect on fishing communities." Supervised by the port authority of Salina Cruz, the construction phase will be limited to less than eighteen months, and any shipping impacts on the lagoon are deemed to be minimal.

Maquarie's infrastructural credentials are also an important feature of the presentation and the company's profile. Promotional materials describe how the company has more than one hundred infrastructure projects, including wind parks in Australia, Europe, Africa, Asia, and North America, and two decades of experience investing in the administration of such projects. The company is reputed, by its own estimation in the presentation, to be globally recognized as a socially responsible company that complies with the highest environmental standards in all its investments. Company press releases underscore these elements again and again.

Mareña Renovables is backed by a consortium of investors with global experience in alternative energy projects. Mareña Renovables is managed by a directing team that is experienced and has worked in all aspects of wind park development, their construction, and operation. Since its beginnings, the primary objective of Mareña Renovables has been to contribute in a positive manner to the local communities and at the same time to provide clean energy to Mexico.

By profiling international credentials such as these, the company may have gained some leverage against concerns about developmental ineptitude. However, flaunting the company's international reputation may have also caused another, unintended effect, namely a more profound link between the project and foreign capital.

On Being Communicationally Sensitive

The park was in danger, and the company had set its focus on enhancing communication with the local communities that would be impacted. In at least one case, they chose an anthropologist for that task. We heard several stories of this anthropologist, employed by the Mexican National Anthropology and Historical Institute in Mexico City (INAH), who was knocking on doors in Santa María del Mar, San Mateo del Mar, and other contentious locales around the isthmus in order to convince residents that the Mareña project would benefit them. In a particularly telling front-page news story on the topic that ran in the statewide newspaper, *Despertar*, it was noted that great efforts were being made to "convince the reluctant Huaves who do not seem to understand modern times."[38]

We encountered an anthropologist in Juchitán one afternoon, but she was not the wind park hawker. However, like the anthropological peddler who had gotten press and inspired rumor, she too was in the business of convincing local communities that the Mareña project was in their interest. Eda had recently been hired by the company as a community relations specialist. Along with another woman employed by Mareña, she was tasked with attempting to mop up the mess. Or at least to nurse communicative relationships back to a point where there might be some hope of a compromise. It appeared as though Mareña Renovables was trying to keep some semblance of cultural and communicational sensitivity intact.

But Eda was worried. She had taken the job, in part, because it was a job. She was recently divorced with a young son, and although her scholarly credentials were laudable (a PhD in cultural anthropology from a US university), she had not found a stable position in her field. "The transition has been hard," she explained to us when we met several weeks later over coffee. It was not fully clear which transition she referred to. Was it having to move from Oaxaca City to the hot and windy world of Juchitán? Maybe it was suddenly being a single mother. Or perhaps it was the job itself, the unenviable task of mending a communication breakdown that seemed beyond repair. It is equally possible that Eda's worries centered around the prospect of losing this job, one that would surely expire if the wind park failed.

As we talked, Eda insisted that we speak in English so that there would be less probability of anyone in the café understanding our discussion. We caught up on what she had been doing in the few weeks since we had first met. She was nervous, occupying her fingers by scooting a paper napkin back and forth on the tabletop as she spoke. She explained that she had been diligently working, along with her supervisor, Edith, visiting the communities in question and speaking with those who were opposed to the park. This was a project of *sensibilización* (sensitization). Its purpose was to let people know precisely what the park would mean for them both at that time and in the future. She described a recent meeting with women in San Dionisio at which they had laid plans for an artisanal workshop where women would be trained to weave palm fronds into hats and baskets, trinkets for tourist consumption.

Eda's job was to talk with people, to convince them, and to coax them. It was probably no accident that two women were sent to perform this kind of communicational labor; their femininity was no doubt enrolled as part of a gentler approach to assuage the public. Media coverage about Mareña had reached a fevered pitch by January 2013. The news was rife with political polarizations, accusations of bribery by the company, and death threats against those opposing the park by what were often described as "bands of armed thugs." The news was almost entirely about the men on both sides of the factional battle over the park. Eda and Edith would give the issue a woman's touch; they would, it was likely hoped, perform a delicate, reparative role in the communities. As Davis had mentioned in his speech in Oaxaca City— and as was very clear in all the conversations we had with anyone in the isthmus at that time—the negotiations were not going well.

According to the company, they had fulfilled their due diligence in speaking with communities. They had conveyed the good works that they would carry out and had informed local populations about the park's impact, both positive and negative. The CDM application, for example, notes a meeting on August 3, 2009, that was held "in order to register the opinion of the people who live in the project zone." Surveys were apparently distributed, specific questions asked, and blank spaces provided for comments or opinions on the project. The company's report states, "All the surveyed people live nearby the project area and they think that it is important and positive that these kinds of projects are developed." Respondents wrote, "This kind of projects [sic] should be developed in the whole zone; That the project proponent should involve people from the zone to increase the labor opportunities; That people from the zone should receive benefits from the project." They believed, according to these statements, that "the lands and the roads should be improved because of the project" and that "this project will bring investment to the zone which will bring benefits to the community."[39] Perhaps most important of all, "It would be good that the project developer organized meetings to explain the implications of the project."

But the implications of the project were also differentially distributed. For the members of the comuna in San Dionisio del Mar who had collective control over the property of the sandbar where most of the turbines would be placed, there was significant financial incentive for the community as a whole to be gained from the park's construction and, ultimately, its profits. The municipality itself had already seen infrastructural improvements in the form of paving and municipal lighting; comuneros—those who were members of the bienes comunales collective—would also enjoy the future disbursement of funds. Across the lagoon from San Dionisio, in the hamlet of Álvaro Obregón, a different set of payments had been initiated for road and transmission line easements; these were distributed individually to private landowners and to the town's coffers. Given the complexity of the different communities affected—communal property holders, private-property owners, populations within municipalities (who were not part of an ejido or comuna), as well as all the governing structures and political factions that accompanied each form of land tenure—it was especially crucial that there be transparency about what benefits or impacts residents could expect. Unfortunately, those lines of communication were either never established or were broken early in the process, becoming further tattered over time.

Although Mareña's representatives noted that the company had held nearly three hundred meetings in the isthmus, in our hundreds of conversations with isthmus residents, we never met anyone who had ever attended one. Many

In reality it will always be a top-bottom approach!

consultations may have been arranged and organized, but it is not clear who was actually there. In the worst-case scenario, only those supporting the local ruling political party and the park were privy to or invited to the meetings. In any case, this was a central reason that Edith and Eda were now hosting regular press conferences and trekking to the farthest reaches of the isthmus in order to perform informational duties that had been left incomplete. In press conferences and conversations with journalists, Eda and Edith were often quizzed and critiqued about the fact that a transparent and robust information campaign was so long in coming. "Why," reporters asked, "did Mareña wait until there was so much trouble and conflict?" Eda was convinced that the company had met its consultational obligations. She explained that in terms of social responsibility, the company was virtuous. "For instance," she said, "we must memorize the ten Equator Principles for the work that we do.... And we've been trying to convey information in the most accessible ways possible, including slideshow presentations or banners, which are easier for people here to understand."

But she also made clear that any social development projects that the company had undertaken or would undertake were not officially required. As one state congressperson had put it to us earlier, for the wind park companies "social development is a *convenio* [agreement] not a *contrato* [contract]." Eda put it more pithily still: "The company is here to generate energy, not to build schools. That's the government's responsibility." And it was here that Eda hit the neoliberal nail on the head. The company was not literally required to provide development enhancements, biopolitical works, or infrastructural augmentations in or around the communities where the park was to be sited. Yet we knew from Davis's discourse that they believed it was imperative to the park's success and duration that the "neighbors" feel "comfortable" with the project. Corporate-sponsored developmental benefits may not have been mandatory, but they were expected. Where resources were thin and governmental projects were often impoverished, these were the sorts of public works that seemed to make privately owned wind power desirable to local residents.

Good Works, Troubled Land

Toward the end of 2012, as part of Mareña's reinvigorated information campaign, attention became focused on how the company had worked or would work to help develop the various local communities that would be

FIGURE 2.1. House, road, electric infrastructure, and turbines, La Ventosa

impacted by the park as it stretched across several sites, villages, and municipal districts. New Mareña webpages appeared during this time, detailing the benefits that residents could expect. The webpages themselves focused on the three communities where protest against the park had already become deeply entrenched. These were also places where divisive politics had often ruled, whether between caciques or, more recently, the national political parties.[40] The roadmap of corporate support for communities indicated in these documents also represented a cartography of troubles regarding land and politics.

In the community of Santa María, Mareña proposed programs to replace school windows, refurbish the *casa comunal* (municipal meeting building), support "cultural" activities, establish capacity-building programs for fishermen, repair bathrooms at the secondary school, and donate an ambulance boat equipped with an oxygen tank and a stretcher. Health brigades made up of a multidisciplinary team would be contracted to foster better health practices in the region. A portable dental unit would be donated and a potable water well installed. In a turn toward the ecological, an environmental education program would be developed.

In the town of Santa María, however, the question of a road was perhaps more pressing. The little town had been blockaded by the neighboring village of San Mateo, preventing anyone from traveling by land from Santa María to the mainland. Instead, as one person described it, "You must go by boat under a hot and smelly tarp across the lagoon." A land dispute over three hundred hectares had fomented bloody confrontations between the neighboring towns over the previous few of years. The standoff continued and was said to have been exacerbated by the wind park contract that Santa María had signed and San Mateo had not. Santa Maríans were receiving payments to hold their land in play for the thirty turbines that were to be placed there by Mareña. And San Mateo, known throughout the region as a "very traditional" ikojts community—a stronghold of ikojts language preservation and usos y costumbres traditional law—was strongly opposed to the park's arrival.[41]

In Álvaro Obregón, as in Santa María, Mareña promised new fences for the secondary school, an environmental education program, and support for health infrastructure programs. The roads would be improved in the little hamlet, and there would be support for "the culture of the Zapotecs"—a particularly ambiguous and undefined gesture, but one that attempted at cultural sensitivity.

Likewise, in San Dionisio, the company promised to refurbish the casa comunal and construct new bathrooms. Support for the development of "the culture" would be provided, and again, health brigades would be sent to the community. A capacity building program for fishermen would be launched. An ice factory would be built. A series of training workshops for artisanal palm weaving would take place, as would the Mareña Cup, a three-month soccer competition. The preschool would have its recreation areas improved, and pedagogical material would be donated to the preschool, the primary school, the secondary school, and the preparatory school. The local basketball court would be restored with new wooden poles, paint, and hoops. They would purchase a new lawnmower for the soccer fields, and balls would be donated all around: soccer, basketball, and volleyball.

As promises were made about the benefits that San Dionisio would enjoy for hosting the park, dozens of people who were opposed to its construction overtook the governmental center of the town and deposed the sitting mayor. The contract that had been signed by the comuneros—back in the halcyon days of 2004—was facing profound legal challenges.[42]

In the beginning, wind power in the isthmus appeared poised to offer nothing but good, all around and in every direction. The "diamond" of the wind would serve as a magnificent and enduring centerpiece to economic and social development. However, the early aspirations for wind power as a renewable resource that would benefit the region and help cure the world's atmosphere were steadily undone. A series of missteps or misrecognitions of how to engage wind power both equitably and transparently, in the places where it has blown so steadily, set the groundwork for future misgivings and a fragile trust.[43] While any contemporary megaproject in Mexico or elsewhere might be prone to criticism, the fact that these dynamics occurred in a place where the state and its agents as well as the mechanisms of transnational capital were already deeply suspect only served to propagate more distrust; these misgivings would deepen, as we will see in chapter 4. Precarious legal and policy regimes regarding wind power development in the isthmus found themselves further debilitated by an ongoing communicational gap between the polished bankers of Mexico City and the isthmus communities that were beginning to see their futures, on the land and on the sea, as endangered. The best efforts of the companies were received unevenly. From the vantage point of many residents, the benefits being afforded to some were either unsatisfactory or too late in coming. Some private-property owners would see rental income from turbines and roads, as would some collective landholders; those who did not possess land would be offered what for many were meager scraps from the corporate table. From the point of the view of the companies, on the other hand, it was never their obligation to take up the infrastructural or social welfare works of the state. Such is the dilemma of neoliberal development, green or otherwise.

A ventifact is an imprint: it bears the erosive effect of sand blown against rock in unceasing wind. The cracks and gaps that appeared across the isthmus where wind power had been set down showed similar signs of wear and the erosion of any trust that may have once existed. But these were ventifacts made entirely by human hands and in the play of suspicions. The paucity of legal and policy infrastructure that characterized the early days of wind power has yet to be resolved. Even in the present, there remains inadequate transparency, community input, and broad access to the benefits of wind power. Outright deception or the dissemination of disinformation may or may not have occurred, but a lack of engaged attention on the part of those who came to build the parks found that the effect was the same. Ultimately, these tensions and suspicions came to rest in worries and conflicts at greater scales: between local ecological impacts and global climatological care, and between sustainable development objectives and concerns about local sovereignty. In the next chapter, we move from the political, economic, and policy attentions of isthmus wind as an infrastructural proposition to focus on an equally consequential but very different material form: the body of the truck.

3. Trucks

Driving in Circles

We heard a story of a man driving around in circles. Teódulo Gallegos, the *agente municipal* for the tiny village of San Dionisio Pueblo Viejo, was the man responsible for representing his community in the deliberations and negotiations surrounding the Mareña Renovables wind park. However, it seemed that he had sold them out for a truck. Pueblo Viejo, situated at the very farthest end of the Barra de Santa Teresa, is a tiny town whose population is ninety-four and declining. In 2012 a reporter for the Oaxacan newspaper *Las Noticias* interviewed Gallegos. At that time, the reporter explained to us, Gallegos was very critical of the park and worried that it would destroy the traditional way of life in his little village. A few weeks after that interview, Gallegos was rumored to have received a gift from the company—a truck—and his mind was changed. He became a strong proponent for the wind park's installation. He was driven, it seemed, by the power of the truck. Pueblo Viejo, because it is situated at the end of the narrow, roadless sandbar, is only readily accessible by skiff or by swimming. "It is too bad," the *Noticias* reporter grinned, raising his eyebrow to foreshadow the facetious comment he was about to expound. "Too bad that they didn't buy him a yacht, or even a little boat instead of that truck." Hindered from ever driving off his little corner of the barra, Gallegos could only ever circle his truck round and round. Looping the village in the wake of the gift, misgivings fell heavily upon him: a traitor for a truck.

IT IS A WELL-KNOWN FACT in the isthmus that the road between La Venta and La Ventosa is perilous for certain trucks. Eighteen-wheel semitrailers risk overturning in winter winds that can reach almost seventy miles an hour. Local newspaper reports are littered with images of these trucks for good reason; it somehow never ceases to shock because it seems unfathomable that such a mass of steel, rubber, and cargo could be felled by mere air in motion. And yet, despite road closures to save them the humiliation and danger of being toppled, eighteen-wheelers continue to capsize and slide across the asphalt of the isthmus highway, their carbonized underbellies prone and vulnerable.

Another sort of truck, however, is more ubiquitous in the isthmus. It is mundane, ordinary, and everywhere: the pickup truck. These kinds of trucks perform a critical role in wind power development, often in very spectacular ways. They serve to move people and to collect them into a shared metallic domain; they also provide the labor of hauling, pulling, and crushing. Rolling unencumbered across roads, trucks may feel like freedom, the inanimate incarnation of human aspirations for progress and prosperity. But in the isthmus, trucks also come to signify people, usually men. Trucks are often seen to be coeval with their owners, and thus, while trucks are venerated and valued, they are also to be beaten and burned. They operate with power and political motive and, at the same time, may embody terror and deceit. But isthmus trucks do not perform a singular representational role or semiotic purpose. They are much more than a thing that represents men or their exploits.

This chapter is a meditation on trucks.[1] It is, perhaps, a (critical) homage to trucks in recognition of their vital instrumentality and their colabors with human actors to produce effects and outcomes. In thinking through trucks as machinic devices that establish and sustain relationships between people, their environments, and the energetic possibilities of the isthmus, my aim is to understand trucks as empowered other-than-human participants: multivalent machines that cannot be reduced to their mechanical, transportational, or representational role. Trucks are instead, I would argue, a node of interrelationship and interchange that, as the feminist theorist Vicki Kirby puts it, allow us to "shuttle across little bridges of translation and transfer"; they enable mutual communication between matter and form.[2] Trucks would seem to be an unlikely nonhuman collaborator in the development of renewable energy. After all, they embody petromodernity in almost every way, from their masculinist stereotyping to their fossil-fueled metabolism.

Trucks would seem to be a survival of petromodernity rather than a signal of its end. However, trucks have everything to do with wind power: they are the machine that drives the process, physically, politically, and often affectively as well. In practical terms, they are, in the literal sense, always at work in building wind farms or transporting the material goods to make them operational. Indeed, trucks could be said to be the most critical machinic actor within wind power, even more than the turbines themselves. It is trucks that capture wind power: in the men they drive, in the politics they create, and in the hopes and terrors they foment. Trucks, I will suggest, operate both as *indicator machines* as well as *transitional objects*—material expressions of human and machinic interplay occupying a space between two ecosocial worlds, one of petromodernity and the other of an aspirational, environmentally viable future.

Indicator Machines and Transitional Objects

In the isthmus, where there is wind power there are also trucks. Both the number and kinds of trucks have multiplied across the region as the wind has boomed and become valuable. Shiny new trucks bought with rental revenue from land that has been leased to wind power companies as well as renewable energy companies' trucks stenciled with their corporate identities—Iberdrola, Acciona Energía, and Gamesa—populate the isthmus in places where they did not before. Fossil-fueled pickup trucks represent an ironic iconicity in the isthmus as they epitomize the continuity of carbon combustion even as the region embarks on unparalleled projects of renewable power. Isthmus trucks thus embody a paradox where renewable energy wealth becomes invested in mechanisms animated by carbon fuel. Wind money buys oily trucks, singularizing a dim line between fossil-fueled modes of modernity and non-carbon-based forms of power.[3]

As trucks in the isthmus occupy a space between petromodernity and a sustainable future, they do the work of what I call an "indicator machine." An indicator, by definition, is not a measure that accounts for quantitative presence. Instead, it is an illustrative example that signals a more generalizable state of being. Indicators reference a thing, a process, a trend, or a movement, but they never presume to explain it. In the domain of biology, an indicator species represents a quality or condition of an environment, a regular and sampleable being that exhibits chemical contamination, biotic disease, or transformed ecological conditions within its body. As indicators, they enable evaluations of biological pasts and probable futures in terms of the present. In a similar way, trucks function as an indicator machine for the Anthropocene. If fossil-fueled trucks would seem to be the epitome of petroculture, I would argue that they are, in fact, much more paradoxical. Trucks that are purchased with wind wealth are also always powered by fossil fuels.[4] Trucks that are used to intimidate opponents of wind power use petroleum to terrorize those people into embracing energy transition. In the isthmus, trucks are not simply metaphors for the carbon age, nor a renewable age in the making. Instead they are at the center of a fraught ecosocial process of transition, an indicator machine. In this chapter, I focus on the ways that trucks codetermine the process of wind power in the isthmus and how they operate within a larger ecology of energy transition. Trucks here are not lifeless matter but rather are forged proof of the resonance between people, machines, and energy forms.[5]

If, as indicator machines, trucks offer a set of commentaries on energy, past and future, they also play a powerful affective role in the politics of wind power, producing dispositions that range from prestige to dread. In this sense, trucks can also be understood as "transitional objects." In the mid-twentieth century, proponents of psychoanalytic object relations were fascinated by "transitional objects" that were thought to facilitate a child's transition from relatively complete dependence (on the mother) to independence (in the wider world).[6] Transitional objects allowed for one's movement from the presymbolic to the symbolic, allowing a child to distinguish a sense of the "me" from the "not me." Transitional objects are not fetishes; they are vehicles for moving from one developmental stage to the next: objects that are betwixt and between.[7] Transitional objects are necessarily transient, never fully part of the self or the other but positioned as a passage between the two: transmutable and impermanent. Such is the status of trucks in the windy isthmus. And, I would argue, such is the status of trucks in an age moving toward renewable futures. They are a vehicle between one stage and the next.

Aspiration and Arrival

You can often tell a truck by what it is not. Cars and bicycles, motorcycles and buses, mototaxis and horses, all have their vehicular roles in the isthmus. But they are not trucks. The mototaxis (or "motos") that swarm the

FIGURE 3.1. Mototaxi, Juchitán de Zaragoza

FIGURE 3.2. Carretón and pickup truck, Álvaro Obregón

streets and alleyways of Juchitán are more ubiquitous and delicate than trucks: soldered metal frames on wheels, fastened to the torso of a motorcycle and tented in plastic to protect passengers from heat and rain. Horses, not trucks, haul trash in Juchitán. Although slower than the motos, it is equine labor that drags drays piled high with refuse, motivated by the long stick in the carriage master's hand. The *carretón* is perhaps the object with a legacy most like the truck. Fashioned from a series of heavy wooden planks and two medieval-looking wooden wheels, carretones have historically been the hauling mechanism of choice in the isthmus; they are nearly indestructible, and they are able to move massive loads of goods, rocks, people, or whatever else needs to be transported from here to there. In a traditional binnizá wedding, it is the carretón that has a central role; the bulky wooden cart is hitched to two oxen and its bed covered in flowers, lilies if possible.

Sitting in the shade of his front porch, Don José believes that carretones— or, better put, the absence of carretones—is evidence of a dramatic change. He says,

> Not that long ago, here in La Ventosa, this whole road used to be carretones. Up and down it. It was dirt back then, and maybe you would see one truck, maybe two on the whole road. But mostly carretones. But now, just look. It is all trucks, all the way down. That is what the wind parks have done. People here in La Ventosa can afford to buy trucks, and they couldn't before. You can see the economic changes here. The road is paved too, so that is good. Because just a few years ago, before the [wind] company paved the roads, these trucks would kick up a lot of dust and dirt. A lot.

For Don José, as for others, the multiplication of trucks is a sign of development and prosperity. They synthesize aspirations for a better life, one with more mobility, more consumptive possibilities, more autonomy, more means to transport family and friends. Trucks provide speed and strength; and they hold a special place in the imagined metrics of progress, and modernity. In such regimes of value, trucks are an objective, a goal, and a sign of one's arrival and economic abundance.[8] In communities like La Ventosa the truck is the accumulation of wealth. Parked outside renovated homes that have gone from being simple single-story structures to elaborate multitiered, multiroomed buildings, trucks are prefigurative of how times are changing and, of course, how the wind has brought new wealth to some landowners.

FIGURE 3.3. Paving the streets of La Ventosa

La Ventosa is a town surrounded by wind parks in every direction. This little hamlet also sits at the crossroads of the two most important terrestrial passages in the isthmus: the Panamericana and Transistmeña Highways. The fleet of Acciona Energía trucks that populate the streets of La Ventosa, parked single file alongside newly installed curbs and paved asphalt, also signal a new environmental and economic era in the town. Wind wealth is invested in new homes and new trucks for some, and the presence of wind development appears everywhere. The infrastructural imprints of wind power are literally underfoot in the form of newly paved streets that the renewable energy corporation, Iberdrola, has partly financed in conjunction with the local government. There is less dust than before. It is a relief to the lungs and a relief from the labors of constantly wiping surfaces covered in *polvo* (dust) carried by *viento* (wind). Critical interpretations of paving also abound, often directed at wind company trucks. Is it possible, some people wonder aloud, that the paving was done only so that company trucks themselves might have smoother passage? Asphalt seemed an uncanny good because, while it may have performed an infrastructural benefit to local residents, it also serviced the needs of foreign trucks and their wind work.

The Drivers

There are two kinds of pickup trucks in the isthmus: those that you drive yourself, and those for which you have a driver. Trucks that have a *chofer* (driver) are invariably white, and they are improbably clean, spotless despite the particulate matter that is part of airborne life in an agricultural and windy part of the world. They are immaculate because the choferes are responsible for keeping them pristine. Choferes always have rags on hand so that while their bosses are inside a restaurant, office building, or meeting of some kind, they can be out with the trucks, polishing their curves. The mayor of Juchitán, Daniel Gurrión, owned such a truck. He also employed a chofer who had an uncanny intuition as to when he ought to add his voice to the conversation in the form of a joke, rejoinder, or observation and when to quietly listen to his employer's speech. The beauty of a truck such as Gurrión's is that it serves as a worksite on wheels, a vessel for negotiations and labor as well as the accrual of political and financial capital. A truck's flatbed can carry two dozen laborers to a construction site, or it might transport an equal number of voting constituents who might need a lift into town. The interior can be made to fit multitudes or simply a handful of people, providing close quarters for intimate conversations out of earshot to anyone except those held in confidence.

Gurrión is a man with a good deal of pride about what his region is capable of contributing to the production of renewable energy. He is a dedicated PRI-ista, and his family name is synonymous with both political prowess and financial might across the istmo. The Gurríons are sometimes referred to as "a family of caciques," a family whose genealogy is rich with powerful local bosses. Gurrión informs us that he has held political office at every governmental level, "from mayor to federal deputy (representative)." He has also practiced dentistry for more than a quarter of a century along the way. Like any man with political powers and persuasive acumen, and perhaps especially like those who are part of the political elite as Gurrión is, there are rumors and tales about the mayor. People speculate that he has ties to the drug cartels and that, although he is married, he is almost certainly queer. What is well established is that his family has significant investments in construction projects around the isthmus and that he is the owner of high-end rental properties, the leases for which are affordable only to wealthy foreigners like Spaniards employed by the wind parks.

As we load into the extended cab of Gurrión's truck, along with two others from Gurrión's entourage, he offers to tour us through several sites where the wind parks reside. We head out of town, and he narrates how squatters have come to occupy huge tracts of land outside Juchitán. Gurrión claims they have all come from an opposing political faction, whom he describes as "the remnants" of the COCEI (La Coalición Obrera, Campesino y Estudiantil del Istmo, the Coalition of Workers, Peasants, and Students of the Isthmus of Tehuantepec). These "camps," he says, are a political ploy for the opposition to literally settle on land owned by private interests so that votes can be swayed and property seized. Groups of people are controlled by various leaders, he says, and when elections roll around or some "political blackmail" needs to be carried out, the squatters are quick to occupy terrain in addition to blocking roads and highways. A look of distaste passes over Gurrión's face. But he does not want to sidetrack our conversation about the parques eólicos by perseverating too long on land disputes and what he views as an unfortunate political modus vivendi embalmed in a state of mutually agreed conflict. His chofer navigates us toward one of our destinations, an overpass out on the highway where you can not only see vast plains of turbines extending for miles on the horizon but also feel the wind nearly blow you over. Gurrión informs us that the wind is now only at maybe half its strength. In the gusts and gales pummeling the overpass, it is difficult to hear. Wind overtakes all human sound. So we retreat to the truck.

In the space hollowed out of the wind by Gurrión's bright-white truck, we hear more of his thoughts. The potential of the wind parks to bring further wealth and development to the istmo is commensurate with Gurrión's general estimation of Juchitán's character. It is a place, he says, that is "*muy comerciante.*" The people of Juchitán, he explains, have "always been merchants and traders. . . . This is the hub of the market region. We understand business." For him, wind parks are the next iteration of this lineage; they are a way to articulate Juchitecos' market mentality with the windy agricultural spaces that surround them.

Gurrión sees trickle-down economics in the wind parks, noting that there is now more wealth in the region, more cars, and (unfortunately, according to him) more mototaxis. But Gurrión is also convinced that better legislation, planning, and care for the history and character of the region and its occupants would have made the installation of the wind parks more felicitous. As he explains to us, "If they had paid better attention, special attention in the beginning of the wind project, if they had thought about the idiosyncrasies of our people, the way they think, the way they believe, the rebelliousness that they have, historically, that they have always demonstrated, if they had paid attention in this sense, to listen to the voices, the opinions," had they done

FIGURE 3.4. The view from Mayor Gurrión's truck,
overpass outside of Juchitán de Zaragoza

this, he concludes, better fortune might have befallen both the companies
and the people. Gurrión describes that when residents of Juchitán, La Venta,
and La Ventosa first learned about the wind parks and the sheer amount of
investment they would bring—in the millions and millions of dollars—they
came to have very high hopes, believing that significant amounts of wealth
would accrue across the community. The failure to meet those expectations
is a sore point for Gurrión. Worse still is the high cost of electricity, which
he gauges to be 1.5 million pesos for public lighting every month for the
municipality of Juchitán. Gurrión is keen on the idea of installing wind turbines that would cover this cost and make lighting free. "By each putting in
a little," he says, "the federal, state, and local governments, with the help of
the wind companies, could make very positive changes here."

From the realm of his chofer-driven truck, Gurrión is able to mobilize
a certain form of political truth, passing squatters right and left, explaining them and disclaiming them. He is master of the interior territory of the
truck; his entourage remains silent as he holds forth and shares his thoughts
and wisdom. As mayor, he is the presumed master of Juchitán, inasmuch as
anyone can "rule" a town renowned for, in Gurrión's own terms, its *rebeldía*

(rebellion). But in the cab of the truck he lives a more intimate political form of communication in motion, moving peripatetically from Juchitán to a wind park, to an overpass, to the next town over, and to the next town after that. It is a bastion from the wind and a place of confidence. The truck is also adept at performing political dominion. Many onlookers know this is Gurrión's truck; its presence is, in this sense, commanding. Making their way into routes and passages, Gurrión and his truck perform publicity and privacy. Observers recognize its personified presence but can only wonder what is being said inside.

An Aging White Nissan

The other type of passenger truck in the isthmus is, in many ways, the opposite of the gleaming late-model vehicle that Gurrión occupies. Rather than chofer driven, these kinds of trucks are steered by their owners, working folk and campesinos. It is this other kind of truck that the resistencia drives into the heart of Mexico City. "It took over twelve hours to get here," explains Tío Sosa, the former agente municipal of San Dionisio del Mar. A busload of protesters caravanned with him and half a dozen other men, traveling the hundreds of miles between the Isthmus of Tehuantepec and the capital city. An aging white Nissan—circa mid-1980s—is their ride. The banged-up little truck has no hubcaps, but it is adorned with rough red-and-black stripes of paint spelling out the acronym of a politically potent teacher's union in Oaxaca, the CNTE 22 (Coordinadora Nacional de Trabajadores de la Educación, section 22).[9]

The little Nissan looks as frail as the trash horses of Juchitán. And like them, it soldiers on, becoming a modular platform of political protest. Carrying members of the resistencia from one place to another, and accompanied by a borrowed school bus filled with women and men from towns and villages around the istmo, the little truck is integral to the protests in Mexico City. The dissent begins on the posh Avenida Reforma, outside the offices of the Interamerican Development Bank (IDB), where the little Nissan is driven as close to the buildings as the concrete barriers will allow. Spilling out of the back of the truck, protesters begin chanting "*fuera*" (out) as they wave hand-lettered signs overhead. Bright-yellow-and-green boards read, "The Isthmus of Tehuantepec Is Not for Sale: Out with the Transnational Turbines," and, "We Demand Respect for Our Sovereignty: Enough with the Plundering." Banners are unfurled bearing similar remonstrations, held in

FIGURE 3.5. White Nissan truck in Mexico City with protestors

each corner by binnizá women in traditional embroidered dresses. Signs and shouts are drawing attention from the well-heeled shoppers and businesspeople on the avenida, some of whom pause to read the messages on the banners that are written in three languages: Spanish, ikojts, and binnizá. The truck, however, is the locus of action.

Two massive speakers and a gas-powered generator fill half the truck's bed, pressing its back tires further into the asphalt. Rodrigo and Alejandro—both longtime activists in the resistance—swap turns with a microphone, and the air vibrates with their demands to speak to those in charge of the Interamerican Development Bank's loans to Mareña. And so the truck goes, loudspeakers blaring words of criticism at each carefully planned stop along the way. Following the sounds emanating from the little truck, the marchers cross several avenues lined with corporate and state interests. In front of the corporate headquarters of Mitsubishi, with curious office workers peering from the windows above, they denounce the company's participation in the Mareña consortium and demand their withdrawal. In front of the headquarters of Coca-Cola/FEMSA, speakers blare similar demands, partly muffled by the coughing rumble of the generator. Finally, we arrive in front of the Consulate of Denmark, which has been targeted because the Danish wind-power company, Vestas, has supplied the turbines for the Mareña project. A representative from the consulate emerges from the building, taking in hand the official letter of complaint that is being delivered at each point in the protest route. Speechmakers have now found a distinct rhythm and tone to their message,

FIGURE 3.6. Protestors resting and listening in Chapultepec Park, Mexico City

rallying the crowd for shouts of support. "Fuera Mareña! No to the pillaging of our land and people of the Isthmus of Tehuantepec." But it is already time to move on. People are hungry, and voices are beginning to go hoarse.

Near the corner where Hegel Street crosses with Tres Picos, we enter the Bosque de Chapultepec, Mexico City's sprawling green retreat. A Mexico City–based organization allied with the resistance has supplied tacos, mandarin oranges, and big bottles of water that are quickly consumed in the shade of the pines. Our next stop, the leadership tells us, is the finale. "Do not lose your energy now!" And soon we are off again, this time farther afield. The truck carries on relentlessly, now to the skyscraper that houses the offices of Vestas and the Macquarie consortium: the money and management behind Mareña.

We cross a good portion of the city to another regal part of the capital. There has been some confusion about the location, but we eventually arrive, and the istmeños are quick to share their dismay at the truly lavish office building. Cased in glass, with a circular drive and a blooming water fountain in front, one could hardly imagine a more appropriate place for transnational capital to hang its hat. Security officers encircle the building, armed and apparently unaffected by the gathering crowd. The generator in the back of the Nissan is geared up again, and the speakers are powered on. The little truck looks especially impoverished in the roundabout of the towering building, and the protestors themselves seem to feel out of place in the bustle of suits and secretaries entering and exiting the lobby. With security in all directions, this is not a building that will be easily penetrated. Rodrigo and Alejandro soon begin negotiations with security officers and their supervisors. A man in uniform starts to jot down names in a tattered notebook, motioning to a handful of people. Our small group will be allowed, finally, to enter the lobby behind the glass façade. "We," Rodrigo proclaims, "are the *comité*."[10] It will be up to us to confront the corporate agents of Vestas, and more important, Jonathan Davis Arzac, the CEO of the Macquarie group and the man behind Mareña. Even once inside the building, however, it is difficult to extract any response. Vestas is refusing to send anyone down to speak with our comité, and they are surely not going to allow us up to their offices. After nearly an hour, the company finally relinquishes, dispatching a short blond man dressed in a suit and tie with a woman at his side. Several of the men in our comité ask for his name and title, but he will not give it.

The man's refusal to state his name is, for Rodrigo, Mariano, and Alejandro, sure evidence of his conceit. For them, this appeared to be an apt reflection of the entire apparatus that they have been battling. With this act of seeming entitlement, the Vestas man now faces a verbal pelting with a new moniker: *prepotente* (arrogant). A smattering of cameras—some belonging to the comité and others held by guards in the lobby—are capturing every moment. Rodrigo, who never lacks verve in clashes such as these, blasts the smaller man, shouting, "We refuse to be exploited in the istmo! We, the indigenous people of Tehuantepec will not have our lands robbed! We are Mexican, and you are foreigners, and foreigners have been stealing from us for over five hundred years." The crowd gathers tightly around the Vestas representative and his female companion, positioning themselves shoulder to shoulder with Rodrigo and shouting, "Viva Tehuantepec!" Rodrigo's finger, now trembling with rage, is raised mere centimeters from the man's face. As a final indictment, Rodrigo spits out the term again: "Prepotente." Signaling to his colleague and mumbling a comment before retreating through the turnstiles of the lobby, the Vestas man disappears. The speaker on the truck outside continues voicing chants, which ring out clear as a bell in the bell jar of the building.

Time is beginning to run out. The security guards appear increasingly restless, and Rodrigo's firebrand spirit and outbursts look like they may soon ebb into violence. Still, there has been no sign of the true object of this mission: a confrontation with Davis. Then we hear a shout. Sergio has spotted Davis coming through the door, probably back from lunch. The CEO is taken aback, no doubt surprised that the protestors are here in Mexico City, and worse still, in his building's lobby. Sergio quickly pops open his laptop and boots up a video in which Davis publicly announced that he would not trample indigenous rights. "But this is what you are doing!" Sergio accuses. Davis has no

FIGURE 3.7. Yellow tape on the facade of the building housing Macquarie
(parent company of Mareña Renovables)

FIGURE 3.8. Protestor with sign, "Like Don Quixote: we are done with your windmills"

comment. With his hand shielding his face, he and his handlers move deftly through the tense crowd, moving bodies out of the way to do so. And then he, too, is gone: through the turnstiles and vanishing into a golden elevator.

Nevertheless, the comité is pleased with the action. All that is left is to paper messages across the building's façade. Someone fetches yellow tape, chosen to resemble crime scene ribbon, and the protestors affix their signs to the plate-glass edifice. Before we are even back in our trucks, security guards are peeling them off.

The little truck in Mexico City delivered on its aspirations. Like Gurrión's white, gliding office in motion, the beat-up Nissan, with its generator and aging concert speakers, performed a certain kind of political act. Its purpose, at least as it rolled along the avenues of Mexico City, was to halt the construction of the Mareña project and slow the growth of neoliberal models of renewable energy production. The work of the little truck was also to abet the men who spoke for and from it. Like the mayor's mobile fortress on the streets of Juchitán, the truck of resistance held political promises, moving bodies and voices to places where they would draw attention to the qualities of energy transition unfolding in the isthmus. In Gurrión's truck, wind power's promise of economic, ecological, and political transformation had potential, but it was not yet fulfilled. In the little Nissan, those same promises faced critique and condemnation. Each testified to the mobile political power of trucks. The gleaming mayoral vehicle rolled through the streets of Juchitán as a political fortress, and the little Nissan in Mexico City offered more insurrectionary potential. In both instances, these trucks can be recognized less by *marca* (brand) than by method; their material form may reflect their placement within a hierarchy of wealth and power in Isthmus, but how each kind of truck is mobilized is testament to the mutable meanings of an indicator machine. How trucks are operationalized depends, of course, on their handlers. But as a venue of political engagement around wind power and a sign of transitioning futures in the isthmus, the truck appears to have few parallels.

Trucks for Treachery

The Restaurante Santa Fe is well known in the istmo as a place where business gets done. As our friend Daniel put it, "Every politician and every businessman has his seat at the Santa Fe." The restaurant lies at a junction of the Pan-American Highway and shares a parking lot with a large and much-used Pemex station. Spending a day or two inside the well-chilled atmosphere of the Santa Fe, it is conceivable that one would come across every important institutional stakeholder in istmo wind power, from congressional representatives and financiers to leaders of the construction unions and turbine engineers. One would most likely not see members of the anti-Mareña resistance inside the rarefied space of the Santa Fe; it is a relatively expensive place to eat, and it is a place where members of the resistance could easily encounter their enemies. Mariano, one of the key spokespeople opposing Mareña, however, is keen to go there with us in order to carry out a covert action, for which two gringo anthropologists will provide the perfect cover.

Mariano is in his twenties, tall, lanky, and sun darkened. His friends in the resistance have always been quick to point out that he is a "good man," one who was once described poignantly as having "simple shoes" and an "old phone." Such descriptions of his material possessions are meant as indicators of his lack of pretension and his humble origins. Mariano lives with and takes care of his mother in Juchitán. We came to know him well in strategy meetings, at street marches, and as he texted communiqués back and forth during tense standoffs. He was being groomed, it seemed, as a next generation of left-leaning political leaders who would continue pressuring the state and corporate actors in the development of istmo wind power. Mariano's reason for wanting to visit the Santa Fe today, he explains, is that he got a text message on his phone earlier in the morning from an unknown number. The text relayed that two comuneros from San Dionisio del Mar would be coming to the Santa Fe to meet with a local PRI official who, at the behest of the company, was going to bribe the comuneros to defect to the company's side. Mariano wants to be there when the meeting takes place. He hopes to catch these compañeros in the act of conspiring with the enemy, even if he might not get close enough to overhear the actual conversation. Sometimes just being seen with the other side is enough to signal betrayal.

Mariano is wearing his usual baseball cap and seats himself with his back to the glass entrance doors of the Santa Fe. He does not want the potentially traitorous comuneros to see him right away. We have arrived separately, at Mariano's request, so that it will appear as though we are meeting him to conduct an interview. "Be sure to have your recorders out on the table," he insists, worried that it might appear that he was being bribed by gringos who might be affiliated with the company. Mariano soon spots the men from San Dionisio. "That compañero" he says, "had been with us [in opposition to the Mareña park] since the beginning. But then he sold out. That is what these other two will do too. They are going to betray us." When the company operative arrives, Mariano has planned that we will use our video camera to record their meeting. If we get lucky in terms of where they sit, we might be able to capture them on tape in their deal making. Or at least this is what Mariano hopes might happen. As we wait for our spy-and-surveillance operation to begin, Mariano checks his watch again and again. He scribbles a quick note and passes it across the table to us. The note says that he saw the San Dionisian men texting on their phones just as they noticed him. No doubt they were planning on changing venues so that they would not be caught. Mariano carefully scans the parking lot in front of the restaurant, looking out to a single row of vehicles just on the other side of the pane-glass windows. He is looking for the briber's truck. He will recognize it immediately, he explains. "But then again," he thinks out loud, "maybe they will just go meet some other place now."

As we sit, anticipating what comes next, Mariano spots another truck, belonging to another Mariano. Mariano Santana has a certain fame in the isthmus, but he is not without blemishes. He led COCEI for a time, only to sell them out to the wind parks, at least according to some. "See that there?" our Mariano says. "That is Santana's truck!" Gifted with the uncanny precision of many istmeños who immediately recognize a man's truck, Mariano is sure of his call. Mariano's suspect never does show up at the Santa Fe that day to meet and potentially bribe the San Dionisians. But Santana's truck will soon appear again. This time, not as transportation but as target.

Encenderlo (Torch It)

The former hacienda of Heliodoro Charis Castro—named for an unlettered fisherman, iguana hunter, and acclaimed general of the Mexican Revolution—had become the home of the resistencia in Álvaro Obregón.[11] In our first encounter with the hacienda, the building was covered with brush and bramble, almost invisible under thriving organic matter.[12] Now, in January 2013, its crumbling brick walls have been cleared of vines, the sandy earthen floor inside swept free of bottles and overgrowth. A single lightbulb hangs at the front entrance and another in the back where the hundreds of people involved in the resistance will gather. Taking turns speaking by the light of the bare, dangling bulb, Alejandro and Mariano are here to share their strategies for the next steps in derailing the Mareña project. Alejandro, for our benefit, speaks in Spanish. Mariano renarrates his compañero's speech in binnizá so that the seventy or so people gathered to hear the plans will be able to understand every detail.

FIGURE 3.9. Meeting by lightbulb at the hacienda; Alejandro
is standing, and Mariano is seated wearing black

An *amparo* (protection order) halting construction on the Mareña park
has just been issued by a judge in Salina Cruz.[13] But Alejandro is concerned
that the judge may fold to pressure. He is clear that Álvaro Obregón cannot
and should not depend solely on legal mechanisms to prevent the Mareña
project from going forward. Moreover, and in a more anarchist vein, Alejandro pronounces, "We are not hopeful that the judge will uphold the amparo.
No. We are not hopeful about that because we are aware of all of the ways
that we are opposed to this grand apparatus called the state." Instead, he
declares, we "need to be *firme*" (steady and strong) in our resolve to stop the
Mareña project "at all costs." Attacks against the resistencia have been coming from all sides, Alejandro elaborates. Most recently, Mareña had begun
"attacking us through the media" (*empezaron a atacarnos mediáticamente*).
The local, regional, and state newspapers, he goes on to say, "have started
to isolate us. They are no longer printing our stories or the interviews that
we, as a movement, give them." Alejandro is compelling, articulate, and the
crowd is transfixed. To underscore the magnitude of the battle at hand, Alejandro reminds his audience that "the interests of big capital are huge, and
they are looking for ways to further increase that capital through this wind
megaproject." Shouts and applause follow in the wake of the speech. But
there is one more important item left on the agenda.

The last point of the meeting is clearly a pressing concern; there had been
murmurs about it before Alejandro had even begun speaking. It is the question

of Santana's truck. Mariano Santana's reputation, at least among the resistencia gathered here in Álvaro Obregón, is not good. He is widely believed to have taken huge amounts of money from Mareña, and he is also thought to have tried to convince others to take bribes and pay-offs as well. Word is that Santana has been seen walking the dirt roads of Álvaro Obregón by night, laden with suitcases full of cash and attempting to persuade landholders to sign on to the project. Alejandro at last turns to the plan that has been quietly circulating: a scheme to send a potent message to Santana that the village of Álvaro Obregón will not be bought: burn his truck. Alejandro guides us through the steps that need to be considered before the vehicle is to be lit on fire.

> We have received information about this truck and where it is. But first we have to be sure, we have to investigate, who is the current owner of the truck. Maybe Mariano Santana gave it to someone or sold it, I don't know. But it would be wrong to destroy it or to burn it if it is [now] owned by someone else. So what we're saying is we need to investigate . . . before we take it or burn it. . . . [In our movement] we need to show more intelligence than those bastards.

Alejandro, who is a master of strategy, has urged caution before; his sense of which weapons and retributions are to be used against state police and others and when to do so is summarily astute. Tonight he is negotiating an exoneration for the truck. But the crowd—especially the younger men—are well poised, maybe even aching, to burn it. Cognizant of the fact that the resistance cannot afford any (more) bad press at the moment, and aware that Álvaro Obregón is often associated with explosive acts of violence, Alejandro proceeds with restraint. "Still," he finishes, "if it is Santana's truck, we are definitely going to torch it!"

Santana's truck, like the ill-gotten truck of the agente municipal driving in circles or other trucks that are, or are believed to be, "gifts" of the company, are loaded with stigma and duplicity. Such trucks are repositories of negative reciprocal relations: a form of cheating and raiding.[14] And for this reason, they have become a principal tactical object that is both vulnerable to attack and representative of betrayal and treachery. Men are known for their trucks and by their trucks. An attack on the truck is therefore an attack on the man. Many at the hacienda that night hope that Santana's truck will burn in effigy. In lieu of the traitor himself, his truck will be made to suffer. If a man's truck can be mutilated in place of the man, it is also true that trucks can become vehicles to instill terror.

Menacing Trucks

There is a whole class of ominous trucks that roam around isthmus towns and villages. They are different from the trucks of hopeful wind farmers, city mayors, protestors, or municipal officials accused of being corrupted by wind wealth and company kickbacks. These trucks are far more dangerous. In the same way that the local press is dotted with stories of overturned eighteen-wheelers in the wind corridor, so too are menacing trucks a regular part of media reports throughout the isthmus. In these stories, the trucks are invariably gleaming white and newer than most. They house ferocious engines and have a capacity for speed and destruction that exceeds the average. In accounts by witnesses or by those who have been the target of their threats, these trucks are typically occupied by "groups of thugs" (*grupos de choque*).

In late November 2012, following weeks of particularly fraught conflict over the Mareña park, one of these menacing white trucks is said to have come looking for Bettina, a binnizá woman and staunch, vocal opponent of the wind parks. On the evening in question, an unidentified man disembarked from his large white truck onto the night-darkened street near Bettina's home. Turning to neighbors sitting nearby, he asked, "Which is Bettina Cruz's house?" The bystanders responded cautiously and, most likely in a refusal to disclose, responded, "We don't know."[15] Frustrated and unapprised, the man mounted the truck once again where he and his unknown passengers sped off into the night.

Trucks like these—white, almost always described as marca Chevrolet— appear like specters in the night or foreboding messages by day. While the occupants are often unseen, the abiding fear that is reported is that there is an enemy inside. Perhaps henchmen sent by the company, or political party operatives. In a setting where those opposed to wind park development have regularly faced death threats, the appearance of the white truck operates like an omen. Its episodic yet ongoing presence functions, as one local reporter described it, as a way of "*intimidando a la población*" (intimidating the population).[16] The big white marca Chevrolet instills anxiety in some and fear in others. In the politically tense times that have been unfolding in the shadows of the turbines, trucks have become instruments of intimidation or proxies for nefarious wishes. They might even be called deadly weapons.

San Dionisio is not a place with much public illumination, and it is not difficult to imagine the darkness of dirt streets at the hour that people would call *muy noche* (very late at night). Indeed, this is a good time and place for murder, or an "assassination," as Isaúl would put it. He and others in the San Dionisian resistance have been occupying the municipal hall for several months, trying to prevent the deposed PRI mayor, Jorge Castellanos, from returning. Castellanos was unseated after having been accused of absconding with millions of pesos that rightfully belonged to the San Dionisio comuna as part of the contract and *uso de suelo* (land use) agreement with Mareña Renovables.

The truck that tried to kill him, Isaúl explains, belongs to one of Castellanos's henchmen. Returning late at night from a general assembly meeting, Isaúl was, as he describes it, nearly run down. The truck was white and had no license plates: a suspicious condition. In its first confrontation with Isaúl and his small group of compañeros on the darkened streets of San Dionisio, the truck began with "incessant honking." The engine was revved loudly, pulsing sound into the night. Undoubtedly, words were exchanged. In a storm of dust, the truck then careened around the men, barely missing them, Isaúl recounts. Circling around moments later at full speed, the truck hurtled straight toward the group of three, braking just in time to avoid running them down. But not in time to avoid hitting Isaúl. With his arm injured—and with a renewed fear of retribution for his opposition to the park and to the local PRI powers—Isaúl filed a report with local police. Isaúl and his comrades testified that the truck was driven by a former comunero who was originally opposed to the wind park. Now, however, he was a "political adviser" for the overthrown mayor who avidly supported the wind park project.

In communiqués from organizations opposed to the wind park, reports of the truck attack describe the incident squarely as "an assassination attempt." So, too, does Isaúl. "They tried to kill me because of my role as a leader in the opposition in San Dionisio del Mar," he explains to us. When we are able to speak with Isaúl's mother, she, too, expresses fear for her son's life. She fully believes that this was murderous vengeance. Local press reports, however, are more neutral in their estimation, calling it "an apparent attack" upon one of the opposition leaders. Days later, the white truck's driver came forward to speak to police. There was no point in hiding his identity; he was already outed by the fact that a man and his truck are virtually one entity. The driver claimed that Isaúl was staggering, drunk, across the road that night and was hit, yes, but not because of any premeditated act of aggression. The truck in

FIGURE 3.10. State police trucks lined up near barricade, Álvaro Obregón

the night that crashed into Isaúl was a potent threat, no matter which account one believes. Indeed, both might have more than a grain of truth to them. Whether or not the truck and driver intended to kill one of the leaders of the opposition is a juridical question, and it may never be clear what the aims and deeds were that night. But white trucks on dark nights invite shadowy accounts. What was the truck's intention that night? The answer undoubtedly lies on which side of the wind park controversy one stands.

In San Dionisio del Mar, as in Juchitán and other isthmus towns, trucks form a particular kind of arsenal. They may operate as a potentially lethal weapon. Or they may make for a convenient way to accuse one's enemies of homicidal intent. In this political calculation, the actancy of the truck is closely bound with the will of its keeper. The resonance between man and machine, and the ways in which both the man and his truck collaborate to craft particular ecologies of fear, suggests that trucks are more than simply tools to be manipulated by human acts. Instead, trucks cocreate the world across which they drive, tuning affective attentions between dread and hope.

The Police

The state is very good at making the presence of its trucks known. Black, well polished, and stenciled with the words POLICÍA ESTATAL, the vehicles of the Oaxacan state police often have a group of five or six officers standing

FIGURE 3.11. Woman at barricade pausing after explaining
what has transpired, Álvaro Obregón

in back, outfitted with bulletproof vests and armed with rifles. They are hard
to miss.

On the warm November morning we first arrive in Álvaro Obregón, we
are greeted by a fleet of state police trucks racing out of town at high speed.
Álvaro Obregón lies on the edge of the Barra de Santa Teresa, and it is the
only terrestrial route to the site of the proposed Mareña park; the barricade
here, which has been guarded day and night by a rotating group of citizens
from Álvaro Obregón and the neighboring village of Emiliano Zapata, has
been a crucial bulwark against the construction of the park. On this morning, we arrive just minutes after a major confrontation between the men
and women on the barricade and a handful of Mareña contractors. In the
hopes that Day of the Dead celebrations would mean a diminished presence
of protesters blocking the barra, the company contractors had made their
move. But they ended up losing, for a time, their trucks.

Minutes after the state police trucks have sped away, we reach the site of
the barricada to find thirty or so people, many of them women and many of
them in tears. We are regaled with a cacophony of reports about what transpired just minutes before. What is clear is that the state police have arrested
eight people who are now en route to the jail in Juchitán for booking. The
crowd gathering around us is very suspicious, and probably rightfully so.

They do not know who we are, and by the looks of us, we could very well be with the company. A few days ago, a team of human rights advocates was here to meet with community members and gather accounts of intimidation or abuse. As we begin to explain that we are anthropologists trying to understand the conflicts around the wind parks, more questions arise. "But, are you with the company?" they want to know. Our taxi driver, Itzail, is nervous about the crowd, which is beginning to enclose his little cab, and he warns us not to get out. "It is too dangerous now," he cautions. But we have already opened the door. Several tearful women speak rapidly in Zapotec, gesturing to the site of the barricade. Itzail, seeing that we were at a communication impasse, pops out of his car and circles around to translate. He, like many people in the region, has Zapotec as his first language and Spanish as his second. "The police were brutal," he conveys for one woman. "She says the police pulled them to the ground by their hair." The woman, who is now tugging on her own hair to demonstrate, is clearly shaken by the encounter. Another woman explains that the police had used "burning" (pepper) spray to disperse the crowd. More people begin to arrive as the news spreads about the confrontation with the cops. Itzail tells us he has overheard that the police had tried to break through the barricade a few days ago and were driven away when the resistance began hurling rocks at them. "The people here aren't afraid," he explains. "They don't have guns, but they have rocks and clubs and machetes and axes." He is convinced, as he confides to us on the ride back into town, that the people of Álvaro are *medio loco* (half crazy). "They have no fear," he avows. "They are insane. They say they'll attack anyone who trespasses, even the police. And they will burn trucks."

Racing back toward Juchitán at full speed, Itzail is thrilled to be driving so fast, with so much urgency, and in the context of so much conflict. Bending around a tight corner of the narrow dirt highway, he barrels over a huge snake crossing the road, crushing it across the middle. Our plan is to notify Bettina about what has transpired and try to discover what has happened to those arrested by the police. As we leave the village, heading toward the highway, we come face-to-face with ten police trucks and four or five other large pickup trucks gathered at the edge of town. Are they awaiting reinforcements? What are they doing here? As the taxi slows to a crawl, we pause to lean out the window and ask what is happening. Appearing to be innocent gringos who are likely lost, we think we might get a response. But the police are stoic, saying nothing. "Just a little problem in the community," they assure us. Itzail hits the gas, peeling away from the scene. "Assholes," he mutters. "They're scared to go in." Flying down the road, we pass more state police trucks on their way to Álvaro. The plot thickens. On the highway we spot two big white pickup trucks parked by the side of the road. Itzail reckons they must be company contractors because one has Jalisco license plates. The men with their trucks appear to be debating their options, waiting for the call. Again Itzail sneers, "They're scared, obviously, to go down there."

Once in Juchitán we set about locating Daniel, our friend who is an attorney. He has been to the jail already, and it seems the *barricadistas* will soon be released. We then go over to Bettina's house and inform her of all that has transpired: the police, the pepper spray, and the near breaking of the barricada. She is quickly on her cell phone, notifying others in the resistance. And we are off again, now with Daniel driving his little green Toyota; this time our trip to Álvaro will be diplomatic.

It is now early afternoon, and the dirt road that leads toward the barra and the barricada is jammed with gleaming black police trucks, threading down the road with officers pacing in their orbit. They are outfitted in helmets and flak jackets, batons and riot shields. Hundreds of residents of Álvaro are milling around the dusty clearing in front of the hacienda. A negotiation has just begun between the police sergeant in charge of the operation and a handful of representatives from the Álvaro resistance. The discussion unfolds as amicable but pointed. They are negotiating the release of two trucks.

With several hundred people standing around them, it is at first difficult to see that, several meters away, the company trucks that tried to enter the barra earlier that morning are now in the possession of the resistance. Both of them have been overturned, taken as hostages. Lying on their sides in the dust and gravel, they too make for a fine barricade. "But how did you do it?" I ask, a little incredulous but also impressed by the group's ability to so fully disable these machines and leave them prone in the dirt. "*A mano*" (by hand), one older man says. "Together," says another with a wry smile.

Eventually, the company contractors would rescue their trucks. A negotiated truce over the course of several days rendered an agreement. It stated that, first, Mareña would desist from entering the barra to begin clearing or construction and, second, that none in the community of Álvaro Obregón would be indicted for the theft of the trucks or any damage the trucks might have incurred. The company representatives very reluctantly added their signatures to those of the community representatives, and a handwritten list of pledges was made. Company agents assured those gathered around that they would refrain from using "state force" to implement the project. And they agreed, finally, to withdraw all their trucks and equipment within twenty-four hours of the signing. The document was captured in a digital

FIGURE 3.12. The resistance with toppled trucks behind them, Álvaro Obregón

photo and then posted, circulated, and archived, with each detail inscribed and digitized. The accord would eventually be broken in a few different ways, a few times over. But the fallen trucks proved to be an excellent medium for drawing the company into a contract that it most certainly never wanted.[17]

The End of the Road

A new wind park is underway in the isthmus, Fuerza y Energía Bií Hioxo, to be financed and built by Gas Natural Fenosa. In Union Hidalgo, Chicapa de Castro, San Dionisio del Mar, and Huamúchil, residents have complained that the heavy trucks used to transport materials for the park's construction have been destroying the highway, their only way in and out of their towns. In grievances to the press and the Roads and Runways Commission of Oaxaca (Caminos y Aeropistas de Oaxaca, CAO), local drivers of cars and pickups have been insisting that they do not want to experience the same fate as the highway outside La Venta, which the construction phase of the wind parks there left "destroyed and [which] remains unrepaired." However, it is more than highways provoking the increasingly vocal and volatile resistance to this newest wind park.

The developer, Fenosa, is no friend of the people, according to many with whom we speak. Occupations had begun at Playa Vicente in order to prevent further entry into the area and to stall the installation of the park. Bettina and her group, as well as Mariano, have been central to this gathering resistance, all taking place not far from where all the Mareña trouble went down. Mariano has recently been jailed after being accused of extortion. The police claim that he profited from the return of trucks that had been appropriated by the anti-Fenosa resistance. Though he was released after a few days, he is required to register his presence at the police station every day until his formal trial. Mariano has since claimed to have been kidnapped and hunted by white trucks on the street where he lives. He believes he is under imminent threat by these malicious vehicles and their occupants. Talk of peril and extortion swirls constantly around the Bií Hioxo park, and trepidation is on the rise just as we are about to depart the isthmus. But there is one last truck at the end of the road.

This truck, like those belonging to the Mareña contractors, has been taken hostage. We have been walking, along with several dozen others, for many kilometers down a long, dusty road. At the terminus, on the outskirts of Juchitán, our group comes to a stop. This is the end of the line. A truck has been taken prisoner. Several men push their collective weight into the metallic body of the disabled truck, slowly maneuvering it across the asphalt. Its owner's identity is clearly marked on the door, a surveying company contracted by Fenosa. Jeers ring out all around, along with shouts to burn the truck. But instead, it will be bashed in. A young man outfitted with a mototaxista shirt ambles up to the truck. With a long, thick, steel pipe gripped in both hands, he smashes it. His fury is palpable. And the crowd loudly cheers him on. As he raises the steel bat high overhead, his feet momentarily leave the ground as he throws the entire weight of his body into the swing. Gravity is briefly undone by his passion to beat the truck. As he slams the pipe against the windshield, the crowd is peppered with flecks of glass. A woman then leans over to whisper something in my ear. "They will not burn the truck," she says. "And why not?" I ask. "Because this way," she says, "they can hit it again another day."

Ending Trucks

Trucks occupy a special place within the ecology of isthmus wind power, helping to form the scope of the possible and to condition the logics of transition. Trucks are a sign of arrival, of wealth and development. They also

FIGURE 3.13. Beating the truck, near Juchitán de Zaragoza

embody betrayal and treachery. Trucks are creatures of violence that can be murderous, lurking in the night, unknown and untamed. And they can be a voice, providing a sanctuary or a platform for politics imagined differently. Trucks and their people co-operate, or as Marisol de la Cadena might say, "co-labor" with one another. Everywhere, one can find trucks that are directly tied to wind power: those owned and run by the wind power companies, those purchased with money gained through land that was leased to wind parks, and those that may have come as a gift to politicians or others who have enabled the growth of wind power. The most dangerous trucks—those that linger and threaten or are overturned, beaten, or burned—are also seen to be allied with the installation of wind power. They exist as an object of desire that might be secured through wind power just as they present an objective threat to those who would oppose it.

Trucks, therefore, can be understood to operate in two ways in the isthmus: as transitional objects and indicator machines. Trucks mediate a space between carbon modernity—which, importantly, has allowed them to thrive—and renewable futures, which, importantly, have depended (at least heretofore) upon them. In this context, trucks here are not inert objects but active participants, moving toward uncertain futures and articulating a relationship between one energetic regime and one that is yet to come. Transitional objects are not fetishes but vehicles for moving from one developmental stage to the next. They are necessarily betwixt and between, a passage between states of being. They are both of a person and not of a person. Of a time and in anticipation of the next. They are deeply ambivalent, affective entities—a bundle of human meaning and material experience erupting in a thing.[18]

Trucks occupy a critical position between the now and the next, as well as between oil and wind. Indeed, even more than the turbines themselves, trucks in the isthmus have proven themselves to be what I came to call indicator machines. As indicator machines, trucks refuse a stable evolution from carbon-powered life to a wind-powered world. They express a process and a movement, even if they do not directly diagnose their causes or effects. Like an indicator species, an indicator machine exemplifies a central truth about an ecology and, most critically, the changes taking place within that ecology. They operate as tools that are ready to hand, facilitating an unknown trajectory from the logic of oil to the logic of sustainability.[19] Trucks thus offer a map toward shifting states of existence. And they are an ironic indicator: carbon fueled and yet channeling the politics of wind power.

If it was wind power that began the saga of transition in the isthmus, then petrol-fueled trucks operate as indicators and transitional objects between contemporary and future forms of energy. The proliferation of the individual passenger vehicle and the Great Acceleration that spurred the quickening of the Anthropocene were commensurate processes. Their genealogies and their carbon legacies are intertwined. Trucks drive themselves into that acceleration—heightening mobility, freedom, control, and power—all hallmarks of that time gearing up to the Anthropocene. But if wind is to the Anthropocene's beginnings as trucks are to the acceleration of it, in the penultimate act it will be species, both human and non, that will come to figure most profoundly in questions of transition and climatic distress. The elemental powers of wind, and the colabors of fossil-fueled machines have deeply conditioned the ecologics of the present, but the animate potential and the vital possibilities of wind's creatures will eventually come to occupy the heart of these questions.

But first, there is a wind park waiting to be undone.

4. Wind Power, Interrupted

At the juncture where green capitalism meets with the human barricade in Álvaro Obregón, dust perforates the air. When the wind blows or a car blunders across earthen roads and rocks, particulate matter seems to enter everywhere. Men in Álvaro are often outfitted to protect themselves from the dirt-filled air: pulling their T-shirts up over their faces to shroud themselves from the airborne earth. Maybe the shielding shirt shows the smiling face of a bygone political candidate, or maybe it is a thinning rock-and-roll concert relic; in either case, the chronic cough that can be heard everywhere in this little hamlet makes it seem as though this is a losing battle. Today, in front of the crumbling brick facade of the abandoned hacienda that the resistance has appropriated as their meeting place, T-shirts have instead been fashioned into masks by a group of young men. Theirs is a more symbolic gesture, signaling a touch of outlaw and an ode to Zapatismo. As they jump down from the bed of a dented pickup truck, even the makeshift T-shirt masks cannot conceal the young men's smiles. They have just returned from an excursion to the site where Mareña has its test tower, a spindly metal steeple with a three-pronged wind vane to meter the quality, duration, and force of the wind. In their hands, the masked men hold something. A prize. The crowd, numbering seventy or so, gathers around, eager to see what the young men have procured. Passing the booty from hand to hand with care, the object finally gets close enough for us to see. It is a gauge of some kind, with numbers on dials and settings and indicators in English.[1] It is an object out of place. "Where did it come from?" we ask. "It fell down from the tower," they reply. "It just *fell* down?" we wonder aloud, a little incredulous.

FIGURE 4.1. Man holding anemometer and wire cutters, Álvaro Obregón

Their smiles growing perceptibly wider, they seem to have decided the bluff was not really worth pursuing. "Well," they say, "it *fell* down when we *pulled* down the Mareña tower."

This is part two of the story of a wind park that never was.

THE MAREÑA PROJECT had a very powerful set of allies and all the forces of transnational capital behind it. But the men in T-shirt masks, along with many others in the isthmus, remained dubious of the park and what it might bring or prohibit in the years to come. As a megaproject with vast dimensions, the park had the potential to affect people's livelihoods in every community bordering its site. It was seen to threaten both the terrestrial environment and the already-vulnerable maritime and lagoonal waters from which istmeños drew subsistence in the form of fish and shrimp. For many, the park epitomized foreign domination. The project, some charged, would result in the expropriation of indigenous peoples' land as well as land owned or collectively managed by local farmers. It would further displace the indigenous ikojts population that had already been exiled for centuries to the watery edges of the isthmus by state neglect and their binnizá neighbors. Government officials and company agents, many claimed, operated through bribes to local leaders who then divided the spoils among their political cronies. Moreover, there were many questions as to whether the park itself was legal at all. A well-circulated claim was that accurate information had not been provided to affected communities. And if the wind park's contract was signed without the signatories' full understanding, then the project was in violation of the Mexican Constitution and international conventions to which the country had agreed. The litany of denunciations against the Mareña project was extensive and widely circulated. And as protests against the park increased, these concerns proliferated in press releases and manifestos targeting media outlets, governmental officials, and international organizations; *denuncios* (denunciations) were posted and reposted on social media. Arguments against the Mareña project operated on several discursive planes, linking environmentalism and human rights, indigenous sovereignty, and the state's obligations to its citizenry.

In this chapter, I follow the lines of protest that gathered against the giant wind park. Over time and in different villages, opposition to the project came under several names, from the *inconformes* (dissenters) in San Dionisio del Mar to APIIDTT (Asamblea de Pueblos Indígenas del Istmo en Defensa de la Tierra y Territorio) headquartered in Juchitán.[2] Those opposed to the wind park assembled, most broadly, under the title of la resistencia or, at times, los antieólicos (literally, the antiwinds). I refer to this process of opposition and its protagonists as the resistance because, as I will show, to claim that those involved are "antiwind" is often a misnomer. What they are instead is opposed to the way in which wind power has come to occupy their lands and seashores in the form of transnational capital, corporate stewardship, potentially spurious contracts, and widespread corruption. The key figures of the resistencia—Bettina, Rodrigo, Alejandro, Mariano, and Jesús—became well known through the Mareña battle, not only in the isthmus but in some cases around the world. With the project's mammoth scale—132 turbines sited across three different communities, each with its specific ethnic, linguistic, and political identification; each with its particular history and land tenure system; and each with its own ecological conditions and concerns— Mareña Renovables faced an almost impossibly complex task. The failure to recognize this complexity earlier or more deeply is one primary reason for the park's demise. The conjoined consequences between place and personhood became vividly clear, however, in the resistance. Here, I focus on the ways in which an array of political and ecological factors acted to interrupt the wind park's formation and, perhaps more importantly, how opposition to the project elicited a more philosophical critique as to whether renewable energy is really anything "new" at all, especially when seen from the point of view of centuries of domination and militant responses to that domination.

What I hope to highlight in the case of Mareña Renovables is that while the company and the state may have believed they were bringing transition to the region—both in the form of renewable energy and in terms of economic development—those opposing the park saw no such transition. In fact, they believed precisely the opposite to be true: the wind park was simply another instantiation of resource extraction and exploitation by outside forces. For the resistance, the installation of the megaproject was yet another neocolonial imposition, robbing indigenous lands for the profit of European financiers. It is not surprising, seen in this light, that the spread of wind parks in the region came to be more generally referred to as *la nueva conquista* (the new conquest). In the interrupted wind park, we see that whether or not transition is, in fact, occurring very much depends upon the space one occupies. Shifting to new forms of energy is crucial to global climate remediation, but these changes will be lived differently across divides of local and global interests. Transition, I will argue, cannot be taken in the abstract; it is not a given, nor is it an objective process. It is, instead, a series of encounters punctuated by particular histories.

Assembling

The office of the Assembly of Indigenous Peoples of the Isthmus in Defense of Land and Territory is readily identifiable on the streets of Juchitán; it is the building with the anti-wind-turbine art stenciled on its facade. As we sit down one Sunday afternoon in early January with two of its founders and a handful of other attentive listeners, it is difficult to ignore our intimate physical proximity in this tiny room decorated with images of past victories and heroes, from Che to el Sub (Subcomandante Marcos of the Zapatista Uprising). Rodrigo begins our conversation and proceeds to detail the historical narrative of the resistance over the course of an hour. Rodrigo, like Bettina (whose story we will hear next), has been one of the primary voices of the opposition; he does not, however, like to be called a leader. This is a designation that he associates with hierarchical, vanguardist, and ultimately corrupt political forms. In fact, for many people in the isthmus, and those who are familiar with its political coordinates, the term "*líder*" is associated with caciques—political bosses who manipulate and control local popula-

FIGURE 4.2. Antiwind stencils on the facade of the
resistencia's office, Juchitán de Zaragoza

FIGURE 4.3. Antiwind stencils on the facade of the
resistencia's office, Juchitán de Zaragoza

FIGURE 4.4. Pictures on the walls of the resistencia's office (Che Guevara and Fidel Castro on left and Che Guevara on right)

tions. It is so often the case, for example, that in press accounts of isthmus politics, the word "líder" is invariably followed by the adjective "*corrupto.*"

Rodrigo is a teacher both by vocation and by nature, which becomes clear in his oration of historical events. The struggle against Mareña, he explains, must to be understood through a longer genealogy that spans many decades and locations.[3] He begins, as many istmeños do, with the region's legacy of defiance against the centralized governmental controls of Oaxaca City and the state-making projects of Mexico City. Like the narratives of local history that we have heard from both taxi drivers and former governors, Rodrigo's account of the region's past emphasizes challenges to any and all external powers. Many people are familiar with the stories from centuries past, prior to the Spanish conquest, which describe how the isthmus Zapotecs served as a bulwark against Aztec invasions.[4] The northern Zapotecs, by contrast, were those who were said to have failed in their self-preservation, allowing themselves to become assimilated both culturally and linguistically. By the nineteenth century, opposition to outside or foreign control was a process of lurching from revolt to revolt: against the central government of Oaxaca in the 1830s and French invaders in the 1860s and then in fervent support of the Mexican Revolution in the early twentieth century. The national revolutionary hero, Benito Juárez, did try to wrest political control over Juchitán and, when he failed, ordered it burned it to the ground. Of Zapotec origin himself, Juárez never receives more than faint praise in the isthmus. Rumor has it that the only place in the entire country where Juárez is truly hated is in the municipal hub of the isthmus: Juchitán.

When the national revolution of independence became locked for decades in the hands of the PRI, who proved to be corrupt and intractable in the second half of the twentieth century, it was the isthmus that responded with rejectionist zeal. It was here that the peasant and student movement COCEI was founded. And it was in Juchitán that the country's first prosocialist and non-PRI-ista municipal government was founded in the 1980s.[5] For local populations, regional histories of defiance were already legendary, but the creation of COCEI saw the area's oppositional fame travel farther afield to garner accolades from leftist revolutionaries around the world.[6] Put in a different register and context, as a former governor of the state of Oaxaca did when we spoke to him in his Mexico City office, "The people of the isthmus are, and have been since the time of the conquest, *muy autónomo y muy guerrero*" (very autonomous and very much warriors).[7]

Rodrigo, however, does not rely on general political histories of the region to render his genealogy of the resistance; he is faithfully citational as he narrates its insurrectionary lineage. Originally, the Juchitán arm of the opposition worked under the name the People's Front in Defense of Land and Territory. However, Rodrigo elaborates, the designation of "frente" (front) is overly encumbered by vanguardism, hierarchical leadership, and military etymology, all qualities they have sought to reject. By consensus it was decided that the nomenclature of "asamblea" better captured their ethos. An assembly evokes, as Rodrigo puts it, "a more indigenous notion, that of community." He goes on to explain, "We have traversed the entire historical process of the Left in Mexico in order to be able to offer an alternative." He narrates parallels between contemporary opposition to wind power and the repression of the student movement in Mexico City in 1968 as well as a guerrilla insurgency in Chihuahua before that. He references the Chiapan rebellion of the Zapatistas following the North American Free Trade Agreement and the battle over the development of another megaproject, the airport in Atenco, just outside Mexico City.[8] He sees the response against wind park development as linked to the teacher's strike in the capital of Oaxaca in 2006 that was guided by APPO (the Popular Assembly of the Peoples of Oaxaca) and finally to Maoism itself with its agrarian peasant insurgencies and challenges to First World imperialism.

Rodrigo's cartography of revolution and response to foreign domination and urban hegemony brings us ultimately to the origins of the resistance against wind parks in 2005. It was founded, he explains, by a group of committed teachers. He and others had protested against the installation of a wind park in La Venta in the mid-1990s, noting that Subcomandante Marcos himself had shown up to speak in solidarity with them. Beyond the symbolic gestures of the Zapatista leadership, Rodrigo continues, the Assembly of Indigenous Peoples could also claim several significant victories of their own. These included nullifying wind park contracts across the isthmus region and, as Rodrigo puts it, "rescuing" twelve hundred hectares of land from being contracted and thus turned into wind parks. Citing potential damage to fishing in the lagoonal region, he, Bettina, and others began helping to convince the very traditionalist ikojts community of San Mateo, for example, to refuse to sign a contract with Preneal in the early 2000s. The qualities and scope of opposition against wind park development in the isthmus have been modular—shifting according to the context, company, and site. Antagonisms toward wind park development have also been contingent upon the level of participatory interest and public reaction to the developments themselves. This is a familiar truth for Rodrigo, who has many years of activist experience. Although antiwind political movements have been ongoing in various forms and degrees of intensity for more than a decade in the isthmus, it was not until 2010, according to Rodrigo, that the conflicts became violent. It was after the apparently accidental shooting death of a man during a melee over a neighboring wind park, Rodrigo avers, that the number of violent threats and acts increased.

Conflicts around wind parks in the region had swollen in scope around this time and, in part, in reaction to the Mareña project. As Alejandro, another founder of the opposition, describes it, "the heart" of the resistance lies across a triangle of communities that would be, potentially, adversely affected by the park's installation; the two key sites of articulation are San Dionisio del Mar and Álvaro Obregón. San Dionisio del Mar is the community that holds collective rights to the sandbar where 102 of the 132 turbines would be located, and Álvaro Obregón is the site of the human blockade on the road that is the only viable terrestrial passage for construction equipment, trucks, and crews to enter the sandbar. From these dyadic centers of opposition follows another series of alliances: local, national, and transnational. The resistance that relayed between Álvaro, San Dionisio, and Juchitán worked with other indigenous rights organizations such as the Mexico City–based AMAP (Alianza Mexicana por la Autodeterminación de los Pueblos) and the northern isthmus organization UCIZONI (Unión de Comunidades Indígenas de la Zona Norte del Istmo).

These collaborations, however, sometimes function as fodder for accusations that an inordinate number of key figures in the antieólico movement are "not from the isthmus." While some of the most prominent spokespeople for the resistance have been reproached for being "outsiders"—or worse yet, "outside agitators"—there is no question that a growing opposition movement was regularly rallying hundreds of people from various locations around the isthmus (and elsewhere) in reaction to the Mareña project.[9] Protestors came from at least a dozen villages, towns, and cities in the region. And they also came from farther afield and through unique coalitions, such as the recently birthed anti-PRI movement #YoSoy 132, launched by university students in Mexico City.[10] Critics of anti-Mareña activism are correct that it is not a wholly autochthonous project but rather one that draws upon partnerships and alliances as well as national and transnational networks.

"Transition"

A handful of compañeros seem to always be seated in the chairs scattered around the courtyard of Bettina's modest home in Juchitán. The aging computer that sits atop a table covered in colorful plastic in that courtyard serves as the ideological portal where all the manifestos, proclamations, and press releases of the Juchitán assembly are crafted, sent, and posted by Bettina or one of her daughters. As a fourteen-year-old girl, Bettina marched through the streets of Juchitán with the COCEI. Since that time, she has been a critical protagonist in nearly every left-leaning political movement in the isthmus. But Bettina is not only a committed activist and spokesperson for binnizá rights, she is, in the opinion of many, the mastermind behind progressive activism in the region. She is credited with making many of the initial pronouncements against green capitalism and the neoliberal qualities of wind parks. Born and raised in the isthmus, Bettina has an incisive sense of local political forms, and she is committed to social justice in the broadest sense. While she began her political life in the COCEI, as that coalition lost legitimacy over the years—through acts of corrupt leaders, bribery, and the loss of its ideological mission—opposition to foreign-backed wind parks became a critical element of Bettina's activist work.[11] But she was also involved in a novel reinvigoration and recommitment to collective, horizontal social struggles.[12] As a key opponent of wind parks in the isthmus, Bettina has also found her political range and voice extended. Her antieólico activism has become world renowned, and Bettina herself has been invited to attend events convened by associations such as the European Parliament and Amnesty International. Her speeches, near and far, hew closely to the overlapping dynamics of neoliberal development projects, foreign capital, and the dispossession of indigenous land.

When we speak with her in August 2012, Bettina focuses on histories of plunder. But perhaps more importantly, she incisively questions the rhetoric and deeds surrounding the call to renewable energy "transition." Bettina wonders aloud whether the form that renewable energy projects have taken could even rightly be designated as a transition at all.

> Maybe we are seeing a transition in the forms of energy, but there is a clear continuity in the form of resource exploitation. These huge companies we have here, sure, they are investing [in the region], but they are taking raw materials without paying for them. They are taking them for free, and everything that they are getting is going to Europe. . . . [These] resources that should be going toward social benefits for people in the region, all of these benefits are going to the multinational corporations. The accumulation of financial resources that these companies have is being used to extract and to exploit people. And it is not just here. It is around everything that they call "energy."
>
> Energy, for example can be petroleum, but for renewable energy, it is now the wind, [it is] solar, geothermal, and biocombustibles, all of these things. . . . So one of the things we question is the fact that it is all the same companies that have plundered the world for millennia and which have now contaminated it. The fact that there is a phenomenon called "global climate change" is because of [their] externalization of costs. . . . [Many of] these same companies have now gotten hold of renewable energy. And so, I have to ask, What "transition"? I don't feel that there is a transition. What change is there? There is no change here. Only talk. And I think that the discourse [about climate change] is being exploited as well. There is worldwide concern about climate change, and the companies are monetizing this as well.

Histories of extraction and exploitation certainly inform Bettina's position. But her attention to the ways that wind energy companies may be or are capitalizing on climatological crisis raises difficult ethical challenges for the future of renewable energy. In her account, the same mechanisms and logics that have driven the fossil economy are here replicated in a putatively noble form of sustainable development. Accusations such as Bettina's have been dangerous in the isthmus. As she and others avow, it is because of her visible role in organizing opposition to wind park developments that she has been thrown in jail, threatened with death, beaten, and driven into hiding.

FIGURE 4.5. Mural in the Isthmus of Tehuantepec depicting Spanish galleons
with transnational wind corporations' insignias on their sails; an armored
soldier to the right holds a wind turbine in place of a staff

Her critics claim that she is simply attempting to leverage her binnizá identity and political history to her own financial advantage. Her many friends
and supporters, however, find her to be an inspirational and committed
spokesperson for the struggle, who is also "one of us." The alliances that Bettina and others have been able to marshal against Mareña have proven to be
formidable. In the wind parks, Bettina has seen herself and her compañeros
as fighting the same exploitative forces that have afflicted the world, and the
isthmus, for centuries, even if it they are now in the guise of wind power.

Disassembling

Jesús keeps a menagerie of creatures in his yard: chickens unearthing insects and bits of grain, a small pig on a rope, and a handsome rooster with
spurs and piercing yellow eyes that occupies a rusted metal cage. Jesus's
wife, Magda, brings around bowls of soup and *totopes* (baked corn tortillas)
wrapped in cloth towels as we gather around a wooden table on their back
stoop. A blank-eyed fish stares up from the bowl of broth, tomatoes floating at its sides; this is a staple around the Laguna Superior. We have spent
the morning talking with Jesús about the protests and the positions of the
inconformes in San Dionisio, the little town where he was born and raised.

It was Jesús's uncle who was the municipal authority in power in 2004, when the original contract for the Mareña wind park was signed with Preneal; but now both he and his aging uncle are squarely opposed to the development. It was in early 2011, Jesús explains to us, that a group of comuneros were able to get hold of the original contract. And this was possible, he notes, only through secret handoffs and a series of clandestine operations. This was the first time, Jesús and others claim, that they were able to view the contract in its entirety, and upon reading it, they were outraged by the project's scope and size. As many of them have explained to us over the months, they were originally informed that the park would be only thirty turbines rather than 132.

In addition to the multiplication of turbines, a fundamental legal error could be traced to a meeting that appears to have never occurred. According to the legal standards of comuna decision making, two meetings must be held regarding any decisions for *resoluciones duros*—that is, decisions involving binding and long-term impacts to collectively held property such as uso de suelo (land use). During these meetings, debate and discussion are imperative, with the understanding that all members of the collective must know precisely the scope and plans of any proposal. Following these discussions, 50 percent of the comuna membership plus one must approve the resolution in order for it to be binding. The inconformes in San Dionisio claim that the second, mandatory meeting was never held back in the early 2000s, and information was not accurately or fully disclosed, and hence, the collection of signatures on the original agreement are null and void. It was, as we would hear many dozens of times, a *contracto leonino*—an unfair contract unjustly favoring only one of the signatory parties, in this case, the company.

During the summer of 2011, when the contract was unearthed, Bettina began meeting with the opposition in San Dionisio, and it was around this time that targeted protests against Mareña began to take shape. In early August members of the opposition in San Dionisio met with other communities on the barra or in the lagoonal region. Although the barra belongs to San Dionisio as communal land—and is legally part of the administrative municipality of Juchitán—traditional use rights, such as the ability to launch fishing canoes from the site, continue to be held by San Mateo, Santa María, Álvaro Obregón, and other hamlets surrounding the sandbar. By midmonth, popular assemblies were being held, and the little village of Pueblo Viejo, in addition to a cohort of San Dionisians, again rejected the construction of the park.

As the year wore on further into the fall, Amnesty International began to receive notices that Bettina was being threatened. By whom it was not clear, but rumor had it that they were henchman working for the company or thugs

FIGURE 4.6. Protest against wind park development, Juchitán de Zaragoza

hired by park supporters who lurked in the shadows or, more often, drove by in menacing trucks.[13] In November 2011 the Office of Human Rights issued a statement to the state of Oaxaca that Bettina required protection owing to the threats (and previous attacks) against her because of her human rights work. Later that month, the Interamerican Development Bank approved the pivotal loan to Macquarie/Mareña, totaling approximately $64 million for the construction of the park. According to an IDB representative, Jeff Easum, the development bank saw this as a move in the right direction and one that fulfilled its green mission and mandate. It was, as he put it in our interview with him, a loan "helping Mexico to take advantage of its abundant wind resources in order to satisfy the growing demand for energy and, at the same time, reduce the importation of fossil fuels for electricity production."

Near Christmastime of that year, Mareña Renovables deposited more than 20 million pesos into the municipal bank account in San Dionisio for the *cambio de uso de suelo* (change of land use) contract. As it later turned out, the municipal president, the PRI mayor Jorge Castellanos, neglected to report this payment, making it seem, to many observers, especially those in opposition to the park, like a very lucrative and ill-gotten "gift."[14]

In January 2012 a growing contingent of comuneros in San Dionisio publicly demanded to "rectify" the 2004 contract. Meanwhile, the mayor announced that he had already signed the construction agreement, just hours before the inconformes' statement. Many residents took this as a unilateral act and one that would only benefit the mayor and his followers both financially and politically. Soon after, the mayor was collectively, and literally, thrown from his office. In an effort to continue the mandate of what Castellanos believed was his rightful political post, he moved his operations to a house down the street. He was, however, effectively deposed and banished from his place of rule in the *palacio municipal* (municipal hall). Several dozen members of the inconformes overtook the palacio, physically occupying the administrative space of the exiled mayor.

On February 8, 2012, a collective of San Dionisians filed a motion in the Oaxacan Congress for official removal of the mayor, claiming that he had committed various crimes and misdemeanors against those he was tasked to serve. Several dozen residents denounced, for example, the fact that they were being denied medical services because of their opposition to the wind park. Or, as one woman put it as we talked to her at her post outside the occupied palacio municipal, "If you are PRD [Party of the Democratic Revolution], no ambulance will come for you. You can forget it! They will just let you die there." Political factionalism has been a part of life in San Dionisio, as in other places in the isthmus, since the arrival of the national political parties.[15] But factionalism had become magnified as the wind park's construction grew nearer.

As the inconformes were taking possession of the town's central governmental space and ejecting their elected leader, the Mareña consortium was expanding as Mitsubishi and PGGM bought into the deal. That same spring, the governor's office was alerted to the trouble around the Mareña site, and representatives of the opposition petitioned the governor's support in their struggle. Our conversations with the governor's upper-level staff indicated that although Governor Cué had not intended for wind development to be a central tenet of his overall legacy, his office was being drawn into the fray and toward the winds of the isthmus. The governor voiced his support for negotiation, mediation, and calm, but above all, he underscored the importance of maintaining capital investment and development in the region. In one of many, many pronunciations in the media regarding his support for the Mareña project, Cué explained, "Given the risk that investors have made in coming to Oaxaca, the obligation of the government, as a facilitator of favorable investment, is to continue dialoging with the groups opposed to the project until they are convinced of the importance of this wind park for the betterment of the planet and for the development of the isthmus region." He warned that allowing this investment to leave Oaxaca would be a very bad precedent.

Across the lagoon, in Álvaro Obregón, the human barricade remained intact. By July, other roadblocks across the Pan-American Highway were being erected by the Obregónian protestors among others. The contention was that Mareña had given 5 million pesos, and an additional 180,000 pesos for "refreshment," to the leadership that had negotiated the electric transmission line contracts in the village.[16] Protestors asserted that this sum should have been allocated equitably among all comuneros, but instead, they said, "It is being divided among only their allies and political supporters." This time it was the PT (Partido del Trabajo, the Labor Party) and COCEI that were being charged with taking bribes. But the ramifications of real or perceived payoffs were the same: more blockades in a place that is quite (in)famous for its ability to mount *bloqueos*.

The governor assured the press and others in August 2012 that despite the ongoing repudiations of the megaproject, the state would ensure the protection of indigenous rights. On the twenty-fifth of August, a comunero in San Dionisio opposed to the park reported being brutally beaten while leaving a meeting of the inconformes. The state government of Oaxaca also stood accused of complicity in acts of violence against the opposition. Meanwhile, in Juchitán, the Assembly of Indigenous People in Defense of the Land Territory continued to roll out missives, publicizing each turn in the battle and elaborating their demands. One manifesto made this call clearly.

> To the wind companies and Mareña Renovable[s], [we demand] that you cease your harassment of the indigenous community of San Dionisio and that you understand that the people will not allow the construction of the park. [We demand] that you stop fomenting an atmosphere of violence and community instability.
>
> To the [Oaxacan state] government of Gabino Cué Monteagudo and the greater state of Mexico, [we demand] that you provide unconditional support for indigenous peoples' right to free and informed consultation and prior consent.[17]

A press release followed a similar line of critique.

> We denounce, once again, [the fact] that we—the people who are being affected by this energy project that benefits only foreign companies—were never informed, consulted nor asked for our consent as is

established in international treaties to which Mexico is a signatory with regards to the rights of indigenous peoples.

In the Declaration of San Dionisio, issued in mid-September, another series of accusations were directed at state functionaries at every level of governance.

> The clear complicity of the federal, state and municipal governments [regarding] the granting of permissions and authorizations that have facilitated and speeded the process of *despojo* [plunder] and death and which have prioritized the production of profit for huge companies' interests has been at the cost of the lives and suffering of entire communities.

Dramatic language of despojo, death, and human rights violations characterized many of the manifestos and statements generated by the anti-Mareña resistance. While their rhetorical contours made rather grave prognostications, these were sentiments that were in accord with many of the reactions of those involved in the debate.

The stakes were high on all sides. The park's potential to polarize communities, disgorge people from their collective land rights, and disturb the waters upon which people have gleaned livelihoods over the centuries all echoed through the narratives of the opposition. Themes of dispossession and cultural survival were woven through these political missives, not only because of the dislocating potential of the megaproject's construction but because many believed that cultural identities would be negatively impacted and worn away over time.

The specific threat of harm to indigenous people in Oaxaca—Mexico's most indigenous state—especially in the context of transnational corporate sponsorship, was an affective turn that was not lost on those who would read and rally around the calls to action made by the opposition. Human rights took a leading discursive role in the claims of the resistencia, but as with many human rights violations and accusations, it was the state that was entreated to repair the damage.

Free, Prior, and Informed Consent

The Mexican state, in its agreement to international treaties, serves as guarantor for the right of indigenous people to free, prior, and informed consent regarding projects affecting them and their lands and territories. Thus, it is the state that is expected to fulfill its obligations and uphold its commitment with regard to potentially harmful development. And yet it is that same state that, in the case of Mareña (and other wind parks), stood accused of complicity in displacing indigenous people and forsaking their rights. The regulatory redeemer is here also the unscrupulous collaborator. Some dispensation is granted to the state, however, in one of the declarations from San Dionisio.[18] The authors write that "Mexico is facing a profound social, political and economic crisis, brought about, in great part, by devastating politics, which have resulted in a series of grave human rights violations against the Indian people of our country." Acknowledging the broader context of financial and political crises, the San Dionisio declaration takes the state as a weak(ened) entity. But this does not condone the ways that economic crisis might have been allowed to devolve into unprincipled deals with corporate actors.

In our conversations with Sinaí, the new *coordinador de energías renovables* for the state of Oaxaca, the Mareña project is "a clear example of how things should not be done." Rattling off the multiple changes in ownership from Preneal to Macquarie to Mitsubishi, Sinaí indicts how the project has been forever "changing actors." He goes on to excoriate the company for failing to contact and be in conversation with his branch of state governance. "Mareña Renovables," he tells us, "has never approached this office in a formal manner. They've never done it!" Speaking in the midst of the Mareña crisis, he is certain that if the company had made contact with his office earlier, or ever, things would not have reached the shrill pitch that they have. But it is more than preemptory management that Sinaí has in mind. Instead, an important role of state government, as Sinaí sees it, is to perform triage and administer ex post facto care in the face of corporate calamity. In the context of the imbroglio in which the company now finds itself, Sinaí offers a sampling of queries and interventions that his office would have put forward had they been asked to do so. "If [Mareña] had approached [us,] the first thing that we would have asked them was, 'What are the worst things that you have done [in the community]? How much coercion has been going on and how much [money] have you been handing out?'" Sinaí is clearly not naïve about the mechanics of corruption and purchased consent in the isthmus. But it is also at this juncture that he sees the strategic deficit of the Mareña project. He continues to hypothetically question company executives: "And did you think that by handing out resources that this was going to make the project *work*?" He mimes passing stacks of bills to them. "It is incredible to have so little faith in the place where you are investing. Trust is earned. And you cannot gain trust with money; you earn trust with a participatory strategy. A strategy where the people know: 'This person is going to bring me benefits.'" Indeed, the benefits that Mareña might bring seemed opaque, and the company's transparency equally so.

A reciprocal sense of distrust infuses Sinaí's narrative. If the company had attempted to bribe (or had actually bribed) municipal leaders or landholders, that action effectively demonstrated that they had no faith in the local populations with whom they interacted; it presumed their corruption and, in turn, playing upon this presupposition, added to the truth of that corruption. An ethos of suspicion seemed to operate unabated wherever we looked, from the sands of the barra to the ignored seats of power in the state capital to the corporate offices in Mexico City. Sinaí admits that he is assessing what went awry from the vantage point of the present and, as he says, "*la visión retrospectiva siempre es veinte-veinte*" (hindsight is always twenty-twenty). But Sinaí is also evaluating the stakes at a moment when the catastrophe is regularly being played, replayed, and played out in media outlets from print to screen. And from this temporal location, Sinaí believes himself, or at least his office, to be innocent of any malfeasance in the process. It might have been different, as Sinaí suggests, if his agency had been involved. Or it might not. Mistrust and accusations of collusion may have simply manifested at yet another level and scale.

Faking Consent

Mareña's legitimacy problem often boiled down to two key issues: the company's apparent failure to properly consult with local residents and the questionable origins of the original contract and its subsequent alterations. In response to accusations of failed consultation, local supporters of the Mareña project devised a plan that they hoped would demonstrate community consent. Whether or not their intentions were noble, the series of community assembly meetings were regularly denounced as "fake." In early September 2012, supporters in San Dionisio, along with the deposed mayor, convened a group of sixty comuneros in an attempt to certify community support for the park through an assembly meeting and vote. Employing a notary public, whom many saw as notoriously unprincipled, the meeting produced a resounding thumbs-up for the project. However, given that a small fraction of the necessary number of comuna members were present, this was a rather weak attempt to simulate community consensus in support of the park. For the resistance, it was a charade. For some media outlets, it was worth reporting on, serving as evidence that the Mareña project was favored by the communities in question and that the park would proceed. As the year advanced, other suspicious comuna assembly meetings were held, with journalists reporting that COCEI leaders were holding fictitious assembly meetings in Álvaro Obregón. In San Dionisio another assembly was proposed, but the inconformes were forbidden from participating. Accusations continued to roll out across the isthmus that people were being petitioned, bullied, and bribed to sign documents that were, in the end, only a ruse.

Castellanos, the deposed PRI mayor of San Dionisio, noted meanwhile that his "patience had worn thin and that vigilante justice was rising." Tensions were running very high. More requests for protection had been made to the Center for Human Rights, stating that key spokespeople were being threatened because of their opposition to wind projects. Carlos Beas, leader of UCIZONI, and Bettina both declared that they were receiving death threats due to their activism against the Mareña project. By early October, the opposition had filed a petition with the Interamerican Development Bank, requesting that the bank cancel its loan agreement with Mareña based upon the company's failure to adequately consult with indigenous communities. Days later, a bus caravan with supporters and foodstuffs for those occupying the municipal palace in San Dionisio was stopped by a bloqueo of PRI-istas and park supporters. The deposed mayor then announced that the governor would have until midmonth to remove the inconformes who had overtaken the municipal building. Or, he proclaimed, he would do it himself by force.

To Congress, and for the Cameras

As conflicts continued to fester in San Dionisio, Sergio and others—in a bid to draw more national attention and publicity to the unfolding debacle in the isthmus—organized a grand caravan of busses to Mexico City. Several dozen istmeños made the trek to the capital city to protest in front of the Mitsubishi building, the Interamerican Development Bank, the Danish Embassy, and finally Mareña itself, where the action reached its dramatic crescendo.[19] A series of news stories followed on the heels of the protests in Mexico City. The left-leaning Mexico City daily La Jornada ran the first story on the demonstrations with the tagline, "Comuneros Accuse Foreign Companies of Wanting to Take Over the Land," noting that if businesses and government entities that the protestors had targeted in their march "failed to stop the Mareña project," they would effectively be "accomplices in the ethnocide of the ikojts people."[20]

The resistance was bringing its message directly to the center of the nation-state, and participants' criticisms were finding bandwidth, just as the organizers had hoped. Toward that objective, the biggest media victory of all was the story that ran on the Dow Jones newswire in English.[21] The article highlighted the financial details of the Mareña project (and the stock market ticker-tape abbreviations of the involved corporations). But it also referenced the critiques of protestors, including quotes from an aging fisherman who was marching: "We are not asking for money. We are asking them to go somewhere else and to allow us our way of life so we can feed our families." On the streets of Mexico City, several people passing by the protests could be heard mumbling accusations of NIMBYism as they read the protestors' banners decrying the wind park. However, the old fisherman's words in print leveled a more basic criticism than complaints about spoiled views or industrialized landscapes that many NIMBY debates comprise. His words spoke to a discourse of survival in a voice of indigenous subsistence that would have a powerful resonance not only in Oaxaca but in Mexico City and across the wires and channels of transnational media.

Vocal protests in the streets and from the bed of a little truck were critical to the action in Mexico City.[22] But the organizers also recognized the opportunity to engage the legislative arm of the federal government. When we next boarded the busses, our group of thirty or so were en route to the Cámara de Diputados, the country's congressional center. Making our way through several layers of security and flashing identity cards and weapon-free handbags at each step, we finally arrived only for much waiting. The regional diputados from the isthmus agreed to meet with the group, but everything slowed to a crawl once we were inside the Cámara. When the congressional representatives arrived, members of the resistance—who had been carefully selected—were asked to tell the lawmakers their stories. The diputados heard from a range of voices, from the leadership to soft-spoken fishermen. Isaúl described how he was run down by a truck; two women from San Mateo narrated their dependence on fish and shrimp and their fear that the park would spell the end of both and thus, their livelihoods. The congressional representatives agreeably nodded their heads and listened. But after the diputados left the room, there was a collective sense that little would come of it.[23]

Soon after meeting with the diputados, our group was shuttled to another room where a press conference was to be held. Sitting adjacent to the

FIGURE 4.7. Press conference in the Cámara de Diputados, Mexico City

congressional gallery with closed-captioned TV screens hovering in each corner of the room, we watched the lawmakers at work in their elegant chamber, recognizing that this momentary proximity to the centers of power could easily slip into utter inaction. In the lobby outside the congressional hall, a dozen tripods were positioned, waiting for video cameras to arrive in order to capture the story of the wind parks and their protestors. Journalists soon crushed in for a ten-minute performance of declarations where a handful of speakers outlined how businesses and government had divided the region into corporate parcels, how the contracts had been signed only under the auspices of misinformation, and how a better model lay in the community-owned proposal for a wind park in Ixtepec, were it only given a chance to go forward.[24] Reporters jotted notes and then quickly decamped. Tripods were swiftly decapitated of their cameras, and soon we, too, were off.

The Deep Fall

By the close of October, stories had been circulating in Álvaro Obregón about a man carrying a satchel stuffed with money to pay off "traitors" to the resistance. Mario Santana, the former COCEI strong man, was said to be at the center of these bribes and perfidious negotiations.[25] Before the month is out, the governor and the president of the Republic, Felipe Calderón, will be attending a grand inauguration of the wind park coming online at Piedra

FIGURE 4.8. President Calderón at inauguration of Piedra Larga
wind park, Isthmus of Tehuantepec

Larga, southwest of La Venta, where the first isthmus wind park was sited. The opposition, though present, is kept far outside the gates by well-armed state police. It is clear that even with official invitations and badges, security is tight. Hovering on the margins of the event, the opposition again declares that according to national agrarian law, any contracts pertaining to land that have not complied with the requirements for free, prior, and informed consent are "illegitimate." At that same moment, President Calderón—hero of climate legislation—takes his place at a podium on the very lands the resistance is naming in their accusations. Calderón makes his case for the power of wind to transform the region and the world. But in the audience today there is not a person who is unaware of the costs and conflicts in doing so. The Mareña project has made the wind more volatile than ever.

As government figures continued to underscore the treasure of the isthmus wind, an increasing exhaustion became evident in the company's pronouncements. For several months, the resistance, too, had been worried about attrition in their numbers. "People are being bought off," Isaúl complained, "and they are getting tired." The threats and constant vigilance were wearing down people's resolve; it had been almost a year that the inconformes had been occupying the palacio municipal, twenty-four hours a day and seven days a week. Concerns were growing about the toll this had taken on the work and family lives of the inconformes. A few days later, in early November, at the Álvaro Obregón barricade, the resolve of the resistance would be tested.

FIGURE 4.9. Protestors outside the inauguration of the Piedra Larga
wind park. Photo by José Arenas.

Taking advantage of the Day of the Dead holiday, when many were
attending to relatives' gravesites or celebrating through the night, a handful of Mareña contractors attempted to break through the barricade to
begin topographical mapping and brush clearing. They were stopped,
however, and their trucks were taken and overturned.[26] A handful of protesters were shipped off to detention in Juchitán, and the state police were
called to arms. "But," as Alejandro explained, "this was when the resistance
was reborn. This was *the* moment that the people understood what was at
stake . . . because people saw that they would not be able to access their
traditional fishing grounds at the *punto del agua*. They saw how they would

not be able to enter the sandbar with the company there."[27] The barra may belong to San Dionisio as communal land, but traditional use rights hold that communities surrounding the sandbar be allowed to launch their skiffs from its sands, as they have for hundreds of years; this access now appeared to be threatened.[28]

After the standoff at the sandbar with company contractors, and after comrades from the resistance have been released from the Juchitán jail, intense negotiations begin outside the dusty space in front of the abandoned hacienda on the edge of Álvaro Obregón. Alejandro stands as the representative voice of the opposition, appropriately clad in a Guy Fawkes T-shirt. At complete ease in the face of authority, Alejandro is crystal clear in his pronouncements.

> After what happened here today, this morning, our position is that we demand the expulsion of these foreigners from our lands and territory. We want absolutely no dealings with this company much less to negotiate with them.
>
> Today, we are firmly resolute that they get out of our town, our land, our sea, our countryside, and that they get out for good [*definitivamente*]. And what we are asking for today from the media, and from the government itself, is the immediate removal of the [state] police force from our town. We are a peaceful people, and this [police] presence we take as a clear provocation. . . . And [you should] understand that this barricade is going to remain, and we are not going to allow even one foreigner to cross it.

The police sergeant in charge of negotiating the release of company trucks and ensuring there is no further violence is equally at ease in the face of opposition, delivering an exceptionally soft-spoken and cooperative oratory to Alejandro and others. The officer asks whether company contractors will be able to leave in peace after retrieving their trucks. Alejandro and others assure him that they will, and that they will be allowed safe passage as long as they are gone by nightfall. "After that," Alejandro notes, "it is a different story." A deal is struck, and the police are off. There seems to be no reason, on either side, to spur further conflict in a situation that has already been aptly described as a *fogata*, or a bonfire.

A few days after the encounter between police, protestors, and contractors at the barricade, company representatives, including Sergio Garza and Edith Avila, are conscripted to speak with the resistance in Álvaro.[29] Alejandro has made it clear that the company must come to them, and that

FIGURE 4.10. Alejandro negotiating with state police sergeant, Álvaro Obregón

FIGURE 4.11. State police officer keeping his camera trained
on our camera, Álvaro Obregón

FIGURE 4.12. Mesa de diálogo abierto,
Álvaro Obregón. Photo by José Arenas.

the deliberations will occur on-site in Álvaro, at the opposition's table, in a
diálogo abierto (open discussion); this will not be a conversation happening behind closed doors or in an executive office suite or in a boardroom in
Mexico City. The community will be present, Alejandro assures us. "But,"
he adds, "we are not going to negotiate. This is *not* a negotiation. We are not
going to make some kind of deal." Concessions will not be tolerated, and
underlying Alejandro's words is the message that they will not be lured by
bribes or promises either.[30] Instead, the Mareña representatives will be, in
Alejandro's words, "forced" to sign an accord.

Several members of the resistance were steadfast that they would make
the ultimate sacrifice for their cause, invoking the famous axiom, "It is better to die on one's feet than to live on one's knees." Any further interference
by the state police or the army, they repeated, would be an act of aggression
for which they would hold the state and company responsible. The secretary
general of the state of Oaxaca would soon be sent on a mission to negotiate
further. He was tasked with convincing the opposition that their interests
would be served and assuring local residents that protecting their rights
was of the utmost concern.[31] Back in San Dionisio later that month, at the

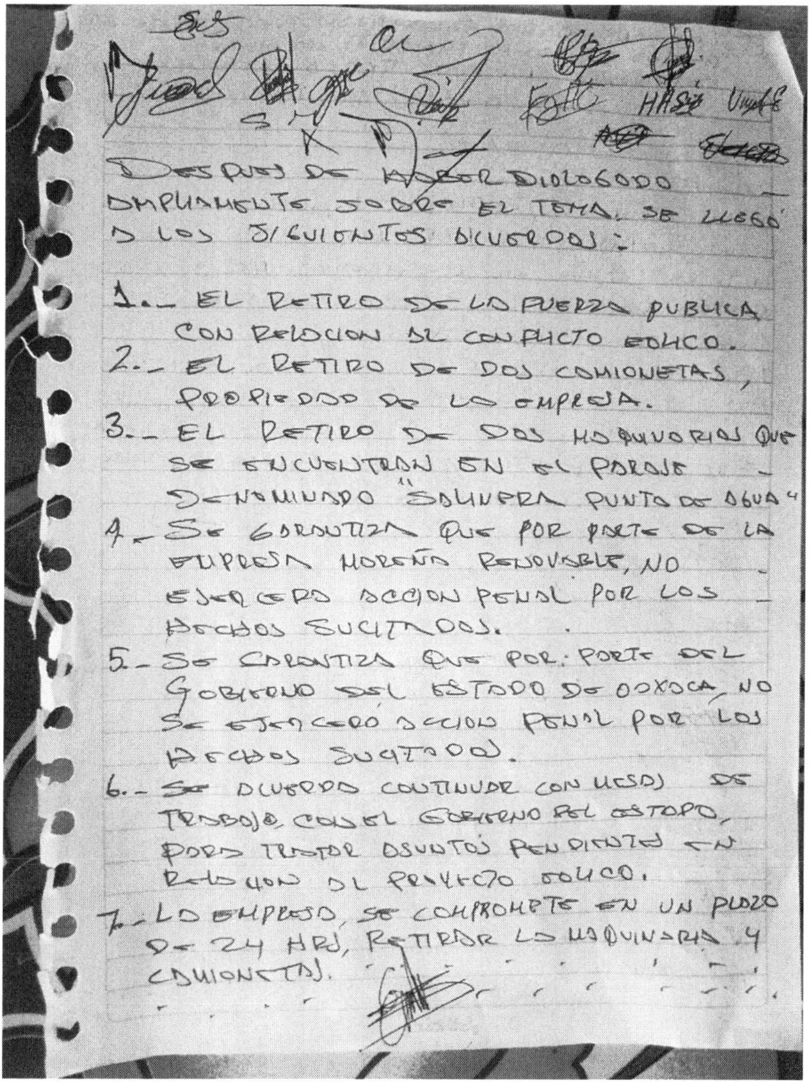

DESPUES DE HABER DIALOGODO AMPLIAMENTE SOBRE EL TEMA, SE LLEGO A LOS SIGUIENTES ACUERDOS:

1.- EL RETIRO DE LA FUERZA PUBLICA CON RELACION AL CONFLICTO EOLICO.

2.- EL RETIRO DE DOS CAMIONETAS, PROPIEDAD DE LA EMPRESA.

3.- EL RETIRO DE DOS HORNOPORIAS QUE SE ENCUENTRAN EN EL PARAJE DENOMINADO "SOLIVERA PUNTO DE AGUA"

4.- SE GARANTIZA QUE POR PARTE DE LA EMPRESA MAREÑA RENOVABLE, NO EJERCERA ACCION PENAL POR LOS HECHOS SUSCITADOS.

5.- SE GARANTIZA QUE POR PARTE DEL GOBIERNO DEL ESTADO DE OAXACA, NO SE EJERCERA ACCION PENAL POR LOS HECHOS SUSCITADOS.

6.- SE ACUERDA CONTINUAR CON MESAS DE TRABAJO CON EL GOBIERNO DEL ESTADO, PARA TRATAR ASUNTOS PENDIENTES EN RELACION AL PROYECTO EOLICO.

7.- LA EMPRESA, SE COMPROMETE EN UN PLAZO DE 24 HRS, RETIRAR LA MAQUINARIA Y CAMIONETAS.

FIGURE 4.13. Stipulations of accord enumerated, Álvaro Obregón.
Photo by José Arenas.

FIGURE 4.14. Signing the accord, Álvaro Obregón. Photo by José Arenas.

occupied municipal building, violence between the opposition and supporters of the park broke out in the early morning hours, and several people
were injured. Violence was coming to be the norm in the battle over the
wind parks. But the resistance's next move was judicial, and it would be surprisingly decisive.

Juridical Object Lesson—The Amparo

It was not media reports, protests, or blockades, nor was it state political
operatives or officials that ultimately spelled the swan song for the Mareña
project. It was instead a federal judge sitting in a humble office in the tatty
port town of Salina Cruz. In the first days of December 2012, a sixty-six-page
single-spaced document made its way to Judge Coronado's desk, outlining
a series of grievances against the Mareña project. The legal entity that the
resistance had submitted was a request for an amparo, a figure of Mexican
constitutional law that serves to protect individuals against injurious acts of
authority through an injunction.[32] Since its origin in the Yucatecan Constitution of 1841, the amparo—an appeal of unconstitutionality—has been used

to challenge governmental institutions and to rectify imbalances between individuals seeking justice and authorities that might prevent them from doing so. A special variety of amparo also exists within Mexican agrarian law that specifically protects the collective welfare of landholders organized in bienes comunales or ejidos.[33]

The document that Judge Coronado paged through claimed that governmental authorities had endangered local populations' rights when they approved the massive wind park project. In authorizing licenses and granting permits, the request claimed, state agencies had deprived community members of their collective right to their land. Moreover, the contracting and implementation of the park violated their constitutional and human rights to free, prior, and informed consent. Similar to a legal challenge that had been filed earlier and that was foundering in the agrarian court, the San Dionisian document insisted that comuneros and other noncomuna members in the region had not been adequately consulted during the original meetings about the project back in 2004.[34] The contract and the project proposal had not accounted for traditional uses of collectively held land, and the original planning had failed to take into consideration customs, social structures, and land tenure systems. Finally, it was argued that the construction and future operation of the project would cause, and would continue causing, environmental and social problems resulting in damage to lands and livelihoods.

The fundamental wrongdoing that the request for the amparo emphasized was that state agencies had failed to ensure free, prior, and informed consent.[35] Additionally, the state had granted licenses based on what appeared to be, at least from the point of view of the complainants, a spurious process. The damage that the amparo was meant to forestall was further erosion of both human and property rights.

> To be deprived in a partial or total manner of one's collective agrarian rights to possession, use and benefit of communally managed lands located on the Barra Santa Teresa [constitutes] a transgression of constitutional rights, the standards of protection of human rights and the rights of indigenous peoples envisioned in the General Constitution of the Republic and in the International treaties [to which Mexico is a signatory].

Signed by several hundred comuneros, in addition to other signatories from the resistance, the amparo was directed to all the governmental agencies that were responsible for allowing these violations to take place. Twelve governmental entities in total—local, state, and federal, from the national Energy Regulatory Commission and the Environmental Protection Agency to the City Council of San Dionisio del Mar—were implicated.[36] Crafting the language and strategy of the document had been a project that spanned from Ixtepec to San Dionisio and from the capital of Oaxaca to Mexico City. The task of the federal judge would be to ascertain whether there was reason to believe that the permits and licenses authorized by the state may in fact have been predicated on the false premises of a flawed contract.

To Judge

At the federal courthouse in Salina Cruz, the security detail consists of three uniformed officers behind portable tables, sheltered from the isthmus sun by a roof overhead. Seeing them perspiring in their once-crisp uniforms, we quickly conclude it will be wise to walk down to the Oxxo convenience store a block away in order to buy cold sodas for both the comuneros sitting on the other side of the gate and the officers themselves. The officers obligingly take the cold drinks, as do the comuneros, seemingly pleased by the offering. Whether petty bribery has any effect is unclear, but after no more than ten minutes, a juridical nanosecond, we are told that the judge would be happy to welcome us.

On the second floor of the building where Judge Coronado has his office, narrow hallways twist off in several directions. Each of them is stacked high with cardboard boxes, brimming with manila folders and files, forming teetering walls that seem especially hazardous in an earthquake-prone part of the world. As we are ushered in, we see that the judge is younger that we have imagined. His desk is tidy but brimming with knickknacks, stacks of paper, and family photos. A portable air conditioning unit murmurs in the corner, keeping the office comfortable in the midday heat. Behind him sits an overflowing bookcase where leather tomes with golden fonts are crushed against paper pamphlets and three-ring binders: a stratigraphy of legal discourse piled nearly to the ceiling.

The judge is very clear that he is more than happy to speak with us but that he cannot share any confidential information about how he will rule on the amparo. His judgment must be kept at bay, he explains, because he is still in the process of receiving evidence. The information that he will be able to impart to our conversation is, he confirms, available to anyone and everyone. Coronado wants to be very clear about his ethics. He also seems very

FIGURE 4.15. Federal court offices, Salina Cruz

intrigued by the task he has been given. The judge claims that he knew nothing about the Mareña case until it arrived on his desk—though it is difficult to imagine that he could have missed the headlines for months. Whatever innocence the judge was bringing to the task would potentially make him all the more *imparcial*. What we hope to learn is exactly how he will approach the case, how he will evaluate it as a juridical object, and if not how he will decide, at least how he seems to be inclined.

Judge Coronado begins by noting that from his legal point of view, this case represents an "extremely interesting issue." Mexican law, he says, provides a "super protection" for community groups, indigenous peoples, and ejidatarios. "Beginning with the Spanish conquest itself," he notes, "certain groups have been subordinated, squeezed, and subjugated [*sobajados*]." At the federal level, he points out, the Mexican Congress has endeavored to protect the rights of vulnerable ethnic groups. And it is here that he sees his role.

> It is up to us, then, as judges, each time we get a case with these characteristics, to be very careful to ensure that these populations are not in danger of losing their natural communal property and that they are not divested from their rightful benefits simply so that someone else may profit. As judges, we must check this since [these populations] can be taken advantage of because they may be ignorant of their rights or entitlements, or because they are not fully integrated into the country, or because they have their customs and traditions, or because they

speak a different language, making it difficult to adequately involve them in the process. Thus, they may not be properly advised. So this is the general framework of the case. And this is what I must ascertain from the evidence presented.

The judge makes it clear that his priority is protecting rights and ensuring that "vulnerable" populations are safe from the exigencies of the state or its corporate allies. His central ethical test here is to assess whether the complainants may have been betrayed by the state, even if inadvertently. As our time with Judge Coronado is coming to a close, he wants to be sure he shares one critical point. "Renewable energy is a good thing for the world," he says. "Of course. But not if it becomes a bad thing for communities." The judge's spare words can, in fact, be taken as diagnostic of any and all energy projects that fail to engage the people and others who are impacted by their presence. His sentiment certainly echoes several years of complaint and protest in the isthmus.

A handful of weeks later Judge Coronado would conclude that there was reason for concern, and he ruled to temporarily halt the Mareña project until a final determination could be made. Headlines rang out with the news that the judge in Salina Cruz had approved suspension of the project in favor of the comuneros of San Dionisio del Mar. His assessment found that government agencies would need to desist from any further permitting, and thus construction, until the matter could be fully adjudicated. His decree read, in part,

> On the basis [of] Article 233 of the Law of Amparo, it is decreed that the acts [of licensing, permitting, and approval for the park] be suspended based on the claims made, to the effect that the relevant authorities do not deprive, in whole or in part, temporarily or permanently, the agricultural goods that belong to the complaining parties, as regards the land located on the barra of Santa Teresa.

Anyone found in violation of the order, the judgment went on, would be reprimanded according to penal law.

Ending the Interruptions

From one vantage point, the Mareña project was a David and Goliath tale of indigenous people struggling for their rights to land and livelihood. From another, the protestors and their allies were merely holding out for political favors, payouts, or revenge against opposing political factions. It may be true that some of those protesting the installation of the Mareña park were waiting for more compensation to come their way, either legally or illegally. Politics in the isthmus are notoriously thick with rivalries and payoffs. But it is also true that there were strong voices in the resistance that rejected the installation of this park and others as a matter of principle: in part because of their megaproject dimensions, in part because of their ties to foreign capital, and in part because of their apparent failure to involve local populations—indigenous and nonindigenous—in the largest-scale industrial transformation the region had ever seen.

While opposition to wind parks had been growing over time in the isthmus, the demise of the Mareña project occurred surprisingly quickly in the time scale of development projects. Multiple attempts to resolve the predicament of the park repeatedly proved themselves to lie at cross purposes: fixes were implemented and interventions made, but they failed to hold. These concerted efforts, eked out over time and endowed with human, financial, and temporal resources, were bound to fail, I have suggested, because they took place in a context already marred by suspicion and betrayal. It became nearly, if not completely, impossible for any of the involved parties to trust the ethical intentions of the others. And yet each was dependent upon those others to achieve their aims. The company believed that it could save its park by placing more money in more hands, by rallying support from state authorities, and through more and better communication. Governmental institutions at all levels were convinced that their further involvement was critical. Adamant voices resounded about the need to preserve the capital investment that the park represented and to ensure an attractive fiscal environment for future renewable energy developments. It was a hard-fought battle on both sides, but in the end, it was a defeat exacted by several hundred committed protestors and ultimately the company's financial hemorrhaging as its turbines lay stalled, indefinitely.

Much of the opposition to Mareña took place on dusty back roads and in blockades of intersections and highways in the Isthmus of Tehuantepec. But the demise of the project also occurred across points of corporate investment—social and fiscal—and through acts of governance, or failures of governance, from the seats of power in Mexico City and Oaxaca City to municipal and communal authorities in towns and villages across the isthmus. The project's denouement also became honed in the portrayals proffered in the local, national, and international press. In this sense, there were many affective and political "atmospheres" at work: from those of company executives committed to green development to those who saw no "transition" at all in the way that new energy forms were occupying indigenous land. Criticism of the Mareña project found political purchase, it is important to note, not because it rejected renewable energy as such. Rather, the opposition codified a suite of concerns ranging from the displacing and destructive potential of megaprojects to worries about the loss of land and fish. And it is toward the latter, the life of species, that the story turned.

5. Species

Runes of Vitality

It is a deceptively simple little diagram. Darwin's famous sketch, drawn years before *On the Origin of Species* was published, maps the variance in reproductive populations: genetic material coaxed in one direction or another, surfacing patterns of mating, birthing, and dying that had previously been vague in their cartography.[1] His is a denuded tree, a rune of vitality, a sketch of biotic cul-de-sacs and pathways moving toward more adaptive forms of life. Even before Darwin, the rhetoric of species and the sciences denoting their distinctions were coupled with "origins"—reproductive beginnings, coming into being, and being a part of a selected collective.[2] As a form of taxonomy, the concept of species has endured since the publication of Carl Linneaus's *Systema Naturae* in 1735. Defined in the most basic of terms, a species is a group of organisms, able to breed with one another over time and in turn retaining an incisive separation from others with whom they do not reproduce. "Species" marks the bright line between deft endogamies and exogamies.[3] But species can also be understood as intergenerational achievements across millions of years and multiple lineages of colabored evolving.[4] The capacious, intertwined logic of species has also placed *Homo sapiens* in the company of other creatures, and in relation with one another, all trying to make what Donna Haraway calls an "earthly living."[5]

Species making began as an eighteenth-century project giving names to the many-creatured and budding environments where humans dwelt. Since the time of Linneaus, 1.2 million species have been found and recorded,

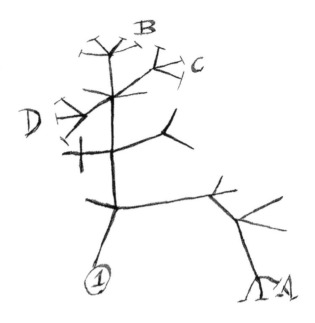

FIGURE 5.1.
Diagram, Darwin's
Notebook B, Transmu-
tation of Species

although recent research suggests that 86 percent of Earth's land species and 91 percent of marine species are still unknown to us.[6] According to some estimates, there are as many "new" species being found and documented today as there were in the mid-1700s.[7] What scientists have now calculated, after much debate, is that 8.7 million animal and plant species currently occupy the planet with us. Human innovations in surveillance and micro-scopic technologies have made new life-forms visible as marine exploration and satellite observation have peered into remote spaces. Lost forest cover and creatures unearthed by the effects of global warming have also been sites of discovery. Out of denuded ecosystems have emerged new species. But more are being lost. A paradox of carbon-driven modernity is that botanists, biologists, and conservation researchers are finding new species in the same moment that the sixth mass extinction is upon us. Estimations of current planet-wide extinction rates range from one hundred to one thousand times the prehuman (or "background") rate. But other calculations paint a much bleaker necroscape, suggesting that extinction rates for terrestrial plants and animals are *at least* 1,000 times higher than the background average, with predictive modeling suggesting future extinctions at 10,000 times the back-ground rate.[8] At this pace, between one-third and two-thirds of all currently living species will disappear sooner than imagined.[9]

In species, we encounter groupings of distinctive life, but we also find extinctions and expirations. Immediate destruction accompanies slow disappearance; abrupt fatalities—like dead birds—populate more gradual demise—such as climate shifts, contamination, and sea level rise—in the "slow violence" of a changed world.[10] Habitat destruction, toxins, and a host of other pressures have been determining factors in the fatalities of species, both plant and animal, since the time of Linnean ordering, but they are accelerating under greater human influence. Just as the term "species" designates tribes of the living—whether wood partridges (*Dendrortyx barbatus*) or amber trees (*Liquidambar styraciflua*), pumas (*Puma concolor*) or humans (*Homo sapiens sapiens*)—it has also been a systematic model to demarcate deathlines.[11] Various flora and fauna have long been categorized as more or less vulnerable, "at risk," "endangered," or susceptible to extinction. Thus, distinctions between species are not just classificatory but an exercise in what kinds of life are to be defended from the contingencies of the present and which will be "set adrift."[12] The Anthropocene hails particular conditions of possibility for species, suggesting new calibrations of vitality and its loss. But it also beckons toward new kinds of colabored care.[13] In the conjunction of new energy infrastructures, multiple and vast extinctions, and well-founded worries about rapidly changing biotic conditions, we need a new reckoning with, and of, species.

Species Being, Species Thinking, and Being with Species

For the Marx of nearly two centuries ago, species was the category of possibility for Man. "Species being" exemplified the ability of Man to transcend his individuality.[14] "Productive life" was, therefore, the life of the species—the collectively produced arts of human ingenuity that Marx believed surpassed all other living things in range, skill, intensity, and adaptability. Capitalism, however, inverted this labor, contorting the life of the species into a means of individual life; rather than thrive in species being, capitalism enslaved and alienated humans from each other and their creations, forcing them into merely physical existence and survival.[15] In our climate-imperiled era, attention centers on a different survival story: that of deep, geologic time that precedes capitalism and its institutions. It has been suggested that this may be a time for "species thinking": explicitly connecting human history to the history of other life. Species thinking might also be a way to reconsider humanity as a form of existence.[16] It is decoupled from political economic systems such as capitalism or socialism; it focuses instead upon the shared boundary parameters of human survival.[17] Species thinking is, therefore, an instrument of being that acknowledges both the consequences of human hubris and the frailty of interlinked life forms. And yet species thinking still relies on a form of human exceptionalism: prioritizing thought produced by humans.

Rather than reverting to older models of species being or prioritizing human cognition through species thinking, what if we were to instead imagine *being with* species? Living as we are in spasms of ecological change, we have an opportunity to reimagine, rework, and refind collective attunement with species. In the present, we (all living entities) are imperiled in ways that we were not before. Therefore, while we humans may have always been a species, we may have only now come to know it. While we might think *as* a species, or think *of* species as a category, establishing a true adjacency with other-than-human life and things is a practice that has been elusive for many, though not all, humans inhabiting the planet for the last several hundred years. *Being with* species would affirm the coincident formation of human lives with other-than-human lives and the material worlds through which we are all mutually composed.[18] Being with species might allow for nonhuman others to differently direct human life.[19] It would be an exercise of recognition because the labor of species would move from being classificatory to consequential: an inductive exercise to surface the mutuality of survival and perishing among and between a much larger "us."

When we stage the survival of species in the Anthropocene and look to sites of renewable energy generation that are meant to reverse that climatological process, we occupy a pivotal moment: a tipping of what Isabelle Stengers describes as "value scales"—the relational balance between human interests for the (so-called) greater good and the suffering inflicted upon other creatures.[20] For Stengers, value scales get revealed in the mix of science, experimentation, and animal testing in scientific labs. Being with species is likewise experimental, recognizing that the Anthropocene itself is a form of animal testing: it involves trials for both human and nonhuman life. Here, I am interested in the value scales of species first as they articulate across spheres of life processes and second in the biopolitical management of species that are, collectively, caught up in the wind.[21] In this chapter, I travel between human articulations of displacement—such as fears about the loss of land and territory—and the ways that humans diagnose, quantify, and seek to manage the species life that is enveloped in wind. I focus empirical attention upon how being with species allows us to see how relational value scales become articulated through "indicators," "conservation importance," and taxonomic practices. Like Stengers, I see these exercises of classification as a particular kind of animal testing, one that uncovers human attempts to establish calculative equilibrium in what might be more accurately described as species roulette.

Creatures

I have never seen as many dead dogs as in Juchitán. This is not meant to be an indictment or to suggest some conscious atrocity—or lack of consciousness—but simply to state a subjective truth. It may be a function of walking a place assiduously, taking the slow time to traverse streets and sidewalks, worn dirt paths and gutters, encountering everything that these places have, seeing what there is to see there. Most of the time, the edges of streets and dusty passageways in Juchitán were filled with the usual rural-cum-urban refuse: crumpled leaves, plastic bags, containers and cartons, spent drinking straws, stones and wrappers of various kinds. But there was also the dead puppy that lay there for days, mouth gaping with tiny teeth. And there was the desiccated dog by the side of the highway leading into town, unrecognizable except for the sculpted skull and matted fur, sinking into the earth and reducing to dust in the isthmus sun. Live dogs with no apparent home or mooring also patrolled the streets, searching for something comestible, sniffing low and wide for anything to fill out the space between ribs and belly. Juchitán was equally home to robust dogs, some guarding homes, others held in the arms of a doting owner in unalloyed companionate affection. The isthmus is not a place where creatures are loved any less, but unlike many urban settings, the lives of nonhuman species feel especially intimate here: trash horses scouring the streets, a cow tearing patches of grass from the highway median, a rooster staring from inside a battered cage, armed with spurs like razors.

Hares—Loss of Land and Territory

This is the kind of animal that would scarcely be missed. Hares are not predators responsible for culling some population of insects, rodents, or other vermin. As sources of prey, they are meager, skinny, and now numbering so few that they would scarcely feed anyone or anything. Those likeliest to feast on them are stray dogs or hungry humans, both of whom would be lucky to even find much less kill this little hare, the Tehuantepec jackrabbit. Wiry and dun colored like its arid home, the Tehuantepec jackrabbit resembles its North American kindred in the southwestern United States. With its slightly transparent, sunset-pink ears cocked and nostrils ready for scents carried across the wind, it does not look remarkable, just another long-legged rabbit on its way to the end of days.

Lepus flavigularis—also known by multiple other names such as the Tehuantepec hare, the Tehuantepec jackrabbit, the Tehuantepec jack rabbit, the tropical hare, the liebre de Tehuantepec, the liebre Tehuana, and the Tehuana hare—was named by a mid-nineteenth-century biologist; he called it the "beautiful-eared jackrabbit." In that time, the liebre was plentiful and surrounded by other creaturely life. According to reports about the region at the time, "The deer, rabbits and hares are innumerable in the Isthmus, the quanitity of the latter which are constatnly passing by the traveller in the plains of the southern coast is almost incredible."[22] In contemporary times the beautiful-eared jackrabbit has been diminishing. The hare has suffered a population decline of more than 50 percent in recent years, more than half its biotic corpus. Their total numbers are less than a thousand and dropping. With no recognized subspecies, the liebre Tehuana is at the end of its taxonomic line. While its waning population would make it seem as though the hare might pass out of existence with little notice, the International Union for the Conservation of Nature (IUCN, Union Internationale pour la Conservation de la Nature), the global authority on species conservation, has deemed the Tehuantepec jackrabbit to be "endangered."[23] Mexican authorities have also designated the liebre Tehuana as "critically endangered."

In its heyday, the liebre Tehuana only ever occupied a little bit of earth, from the city of Tonalá in Chiapas to Salina Cruz on the coast of the Oaxacan isthmus. But its hopping grounds are now limited to small patches of savannah; the remaining populations of hares are rarely found, and the groups are isolated from each other. Their range of habitat and the quality of that habitat have been reduced by the human impact of agriculture, industrial development, and pollutants, both organic and inorganic. Those forces continue unabated. To make its way in the world, the hare now occupies only a fraction of its original territory, somewhere between sixty-seven and one hundred square kilometers. Four small, separate, and separated popula-

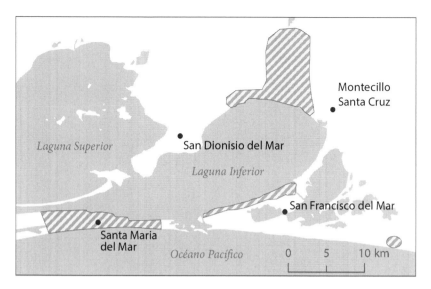

MAP 5.1. Habitat zones of *Lepus flavigularis*. Created by Jean Aroom with assistance from Jackson Stiles and Hannah Krusleski.

tions make a patchwork of hare habitat—one of which is on the peninsular stretch of land where the town of Santa María is located and where thirty of Mareña's 132 turbines were to be built.

The liebre of Tehuantepec prefers to live no higher than five hundred meters above the sea in grasslands with both open shrubs and trees for cover. Sparse bushes—*Byrsonima crassifolia, Opuntia decumbens, Opuntia tehuantepecana*—and scattered trees like *Crescentia* (the *jicaro* tree) are its vegetation community. *Sabal mexicana* (palmetto trees) thrive along the coast and near the saltwater lagoons of the Gulf of Tehuantepec's northern side. Savannas and coastal dunes dominated by *Paspalum* and *Bouteloua*, locally known simply as *grama*, provide critical protection for the hare. Like *Sylvilagus floridanus*, the eastern cottontail with which it shares these environments, the jackrabbit is crepuscular and nocturnal, preferring the shadows of twilight or the full cover of night sky.

Despite its wisely chosen habit of night living and the presence of friendly vegetational cover, humans have had their way with the hare. The jackrabbit's "use and trade" designation notes, "This species is hunted locally for subsistence and very occasionally taken as a pet." The hare's precarity is a carefully detailed account, drawing out the relational links between jackrabbits and

FIGURE 5.2. *Lepus flavigularis*, the Tehuantepec jackrabbit. Illustration by Mario Norton.

TEHUANTEPEC JACKRABBIT
(LEPUS FLAVIGULARIS)

the large primates in their range. The report notes encroachment by agriculture as the human population expands, detailing that "the Tehuantepec jackrabbit is jeopardized by habitat alteration and degradation by introduction of exotic grasses, human-induced fires, agriculture, cattle-raising activities, and human settlements in savannas." The rabbits are also killed by hunters who come from cities up to 200 kilometers away to shoot deer by spotlighting at night. Predation, human-induced fires in the savanna, and poaching all afflict the liebre as well. When coupled with low genetic variation and habitat reduction, the hare has few places to thrive, ecosystemically or biologically.

While the Tehuantepec jackrabbit is listed as critically endangered according to the official standards of the Mexican state, the IUCN report states that "conservation laws are not enforced by the local authorities." In spite of its precarious status, the liebre does not, according to the IUCN, "benefit from protections." Facing fires and feral dogs as well as humans with shotguns and a hunger for land, the Tehuana jackrabbit had been in the process of being killed, burned, and occupied out of existence. The arrival of wind power posed yet another threat to a deracinated species on the edge of extinction.

In the classificatory logic of species, the "environment" takes on a diagnostic subjectivity, either ailing or healing. The Interamerican Development Bank has relatively strict regulations about the environmental and social impacts for development projects that it funds, such as the Mareña wind park. In Mareña's proposal, the barra and the region around Santa María were depicted as being "severely affected by anthropogenic activity," with urbanization, agriculture, small-scale salt extraction, and the grazing of cattle all contributing to the region's changed ecological conditions. By contrast, a mid-nineteenth-century team of Spanish surveyors found this area to be "clothed in a luxuriant vegetation somewhat resembling that of the parks of Europe."[24] As they described it, "The flowers in some of these localities are of astonishing beauty."[25] In the long interregnum between surveyers combing the region to assess its utility for a transisthmus canal and the installation of wind turbines, much biotic change had ensued even as the logics of foreign prospecting remained similar.

The tracts of land where wind turbines were to be placed, were also home to deciduous tropical forest, spiny forest, halophyte vegetation, aquatic and subaquatic vegetation, mangroves, and sandy dunes. Indigenous plant species like *Byrsonima crassifolia* bushes, locally dubbed *nanchal*, form clusters of sparse shrubbery. A grass-forb understory—comprising herbaceous flowering plants, or *morro*—are a critical element of the Tehuana hares' primary, core habitat. According to habitat-use research carried out in Santa María at three junctures over the course of one year, nanchal and morro are crucial to the liebre Tehuana's survival. The understory provides vegetal cover so that the hares may feed and breed, rest and rear their leverets, detect predators and evade attacks. Originally, Mareña's turbines were to be on the western side of Santa María, amid the scrub grassland, but a new siting protocol was devised in order to avoid trampling what was designated as "critical vegetal infrastructure" for the jackrabbit. It had been decided that the park would be moved to protect the animal's home. The liebre had thus been rendered as a medium of what Fabiana Li calls, "equivalence." The rabbit became a site for bodily quantification and a metric for expert intervention.[26]

Environmental Impact

The building housing SEMARNAT (the Mexican Secretary of Environmental and Natural Resources) is filled with managers, overseers, and administrators whose job it is to ensure adherence to the laws governing environmental resources. Located in a prosperous commercial neighborhood in Mexico City, the edifice itself is an homage to high, concrete modernity. The elevator system, conversely, is a page out of magical realism, stopping only on every fourth floor. Alberto Villa is the director of evaluation for the energy and industrial sectors, and he has an office set high in the scheme of SEMARNAT. From his felt-covered conference table, the capital spreads out in hazy horizons in all directions. We are late to our meeting with Alberto, lost in a maze of hallways that seem to defy logic. But he is welcoming and warm. Alberto pulls out a massive tome authored by SEMARNAT to ensure that legal obedience to the nation's environmental laws and resource regulations are carefully and assiduously followed. With the volume at his side, Alberto begins to explain the intricacies of protections. In SEMARNAT, species are subject to the human machinery of bureaucracy, teetering between the protocols of oversight and care and the inertia of institutional practice.

There are manifold layers of governance and bureaucratic will that animate environmental adherence. The mandate of SEMARNAT, as Alberto makes clear, is to test the legitimacy of a proposal against the requirements of Mexican law. "We are not scientists," he reminds us. Officials at SEMARNAT do not determine whether any project scenario and its accompanying environmental impact studies are correct or false but rather whether the development follows the contours of legislation that have been established in policy.[27] "To build a wind park," Alberto begins, "you must start with SENER [the secretary of energy] and CFE [the Federal Electricity Commission] and be granted a concession." The second stage requires that the developer secure environmental impact studies and report each of these conditions to SEMARNAT for evaluation to, ultimately, obtain the requisite environmental permits. There are several stages and kinds of environmental credentials to be acquired—the *cambio de uso de suelo forestal*, a permission for "atmospheric impact," a cambio de uso de suelo at the municipal level to begin construction, and hazardous waste management and disposal. But the first step is to pass the test of SEMARNAT, beginning with the *manifestación de impacto ambiental* (MIA, the environmental impact assessment).

The manifestación de impacto ambiental is a document contracted by the developer and executed by third-party consultants. The manifestación describes the proposed project—its dimensions and duration, materials and magnitude. It must evaluate and speculate on a proposed project's potential impacts as well as propose mitigation and prevention measures.[28] It additionally needs to situate each element of the proposed infrastructure within an "environmental system." The current state of "degradation" as well as the "environmental services" (*servicios ambientales*) provided by "the system" must be described and enumerated.[29] If the baseline of the manifestación is to reflect what is environmentally extant and what kinds of injury currently prevail, the true purpose of the manifestación is predictive. It is poised to proffer a future and to be an exercise in scenario making and risk attenuation. It is a prognostication about human-imposed future damage and a detailed portrait of the expected outcomes. The manifestación is a prospecting device with an anticipatory method.

Alberto, wanting to be clear about SEMARNAT's regulatory schema, details how each scenario must be crafted. First, the manifestación must determine what would occur in the long and medium terms if the project were *not* constructed. This is a baseline of probable "continued deterioration."[30] Second, what would the predicted future hold if the project were developed without any preventative measures for the environmental system? And last, a final scenario must document the projected impact of the project if preventative procedures were put into place. These criteria of environmental impact are commonplace, a widely accepted norm of international environmental law—pronouncements and guidelines that are intended to balance economic benefits against damage to ecosystems and certain species. Mexico's environmental assessment process, like many across the world, coincides with widely agreed-upon norms and conditions set by bodies such as the United Nations.[31] Enacting them is not always ensured, however.[32]

For the manifestaciones that are processed through government agencies, all possible environmental damage or projected impact of an infrastructural project must be addressed. If water use is implicated (for hydropower production, for example), then other state entities must be tapped, such as CONAGUA (the Water Commission). In the case of impacted forests, the General Law on Sustainable Forestry Development must be followed. The same holds true for wildlife or endangered species under the auspices of PROFEPA (the federal attorney for environmental protection). At all levels of Mexican governance—national, regional, and local—ecological planning devices are meant to provide surveillance over urban growth as well as agricultural and industrial land use.[33] Mexico's ratification of the Declaration of the United Nations Rio de Janeiro Summit in 1992 bolstered this stance, committing the state to "develop national legislation regarding liability" in order to compensate those who are "victims of pollution and other environmental damages." Equally important is the amendment to article 4 of the Mexican Constitution, which asserts "the right to a healthy environment." Each of these legislative commitments presumes that "victims" of pollution are human and that those who have the right to a healthy environment are human as well, even if these risks are to be attenuated at the levels of nature and the environment.

Legislative tiers that interconnect the lives of species, are intended to perform the work of *sobre posición jurídica*—a set of critical legal redundancies—in order to provide safeguards for the environmental system implicated. Governmental "partners," Alberto explains, are very important to the process. "And," he notes, "these [government] partners usually, as a rule, assent only to conditional approval for a project." They are prone to argue for more intervention, not less, before agreeing. "Vigilance is important," Alberto stresses, because, as he pointedly remarks, "sometimes the local impact [of a project] is much more environmentally detrimental than climate change is." Alberto's declaration uncovers a point of precarious tension, where localized impacts are to be endured in order to ensure translocal climate remediation and greenhouse gas reduction. This depends, of course, on how environmental detriments are being scaled and their value accounted: for the benefit of one or another species or (as Isabelle Stengers might pose it) for the "greater good."

Culminating the manifestación consultation process is a large public meeting where everyone in the community, as well as experts, are invited to participate and have the project presented in what Alberto calls "a personal form." This gives people, he says, "a chance to let off steam all in one day." As with all projects of scale that pose threats to ecosystems and human health and well-being, there is a need to actively "deliberate discord" among the affected publics, human and nonhuman. It is no coincidence that in the heat of such discord, Alberto would draw attention to the blowing off of steam, underlining the affective stresses that these encounters produce.

Alberto is also clear that while SEMARNAT is open to expert and community commentaries, the office would not take complaints in the form of someone saying, as he put it, "I don't want this wind park near me because I don't want it." (This is what Alberto might have called the whine of NIMBY.) Instead, critics must have an *environmentally sound* argument; it must be predicated on verifiable ecological umbrage, and all concerns and critiques must be submitted in writing, a mode of complaint typically reserved for the educated. A provisional approval can follow at this point in the review, but any impacts that have been forgotten in the original manifestación, or that were inadequately addressed, must be resolved in order to acquire the provisional status of "environmentally viable."[34]

Conditionality provides important leveraging throughout this process. Officials at SEMARNAT claim that projects are never approved in the first go-round. There is always more to be done. For Alberto, the consultation process is a way of "double checking" on the company. He describes it as a brainstorming procedure—*una lluvia de ideas* (a rain of ideas)—a negotiated play of ratifying, extending, and correcting. "There may be more work to do," Alberto explains. "We may not sign off on the environmental impact reports, and more studies may be required or attention given to [a particular plant or animal species] before we will grant the permit."[35] If the studies prepared by the developer fail to discuss in detail a protected species, for example, it is rejected and sent back. As Alberto puts it, "Try again."[36]

Cut and Paste

Saul Ramírez is someone who ably navigates the details of environmental impact reports. Our conversation with him in June 2013 took place on Skype, crossing the space between Mexico City and Oaxaca City where Saul's company, Gestion Ambiental Omega (GAO), is located. Animated on the screen, and sometimes frozen midstream, we conversed with Saul and his colleagues, Edith and Magda, about the procedures of environmental impact reporting. GAO produces reports to be submitted to SEMARNAT, CONAGUA, and PROFEPA among others to evaluate the many dimensions, contexts, and *medios* of environmental spaces and species that might be impacted by proposed projects. The would-be developers contract and finance any field investigations, archival and scholarly research, and report production by organizations such as GAO.

"To build a wind park," Saul begins, "you need two key permissions": an environmental impact permit (from SEMARNAT) and a cambio de uso de suelo (change of land use) permit that attends to effects on "native vegetation," especially forests.[37] The first is a description of the project and its physical footprint, including roads and turbine siting. The second is a discussion of interrelated "mediums," the *medio físico* (the physical setting and context) and the *medio biótico* (the biotic setting). In the case of wind developments in the isthmus, Saul notes, these mediums are lagoons, forests, and (human) communities, including what he calls "their customs and population demographics." The biotic setting, in good species-tracking form, includes human populations and their quantifiable "demographic presence." But attached to this biotic medium are also their "customs," their culture.[38] The presence of protected or endangered species, Saul points out, represents a definitive reason to tell the company to "move the road, move the substation, move whatever." In these cases, Saul notes that it is their ethical and legal obligation to consider shifting the coordinates of a project. And it is, in fact, toward these sorts of ethical questions that we have been slowly moving by walking through the many details of calculating, measuring, and validating species so that new energy infrastructures may be permitted.

A bit tentatively, but with great curiosity, we raise the issue of rumors that we have heard. Some have said that environmental impact reports have been, as one journalist put it, "cut and pasted" from other studies in other regions; such reports are essentially bought and paid for. But Saul rejects such speculation. "That would be unethical," he declares. "And in any case, the legislation is very clear that the data must be specific to the project site." But he pauses for a moment and reflects. "Pirating" studies might happen, he admits, but it is utterly unethical. Besides, he says, it would be quite difficult to "put one over on SEMARNAT."

My Lagoon and Our World

The bureaucrats of SEMARNAT or private report makers like Saul expertly engage in species accounting. They appear to take their work very seriously, both observing the mandates of governmentality and law as well as perceiving the environmental and creaturely needs that they are tasked with surveilling, protecting, classifying, and aligning with (human) priorities. As Kathryn Yusoff might see it, practices of attuned sensations and new sensibilities become compulsory in encounters with "diffuse, recalcitrant, and dislocated" issues of biodiversity loss, new biotechnological life, and of course, a changing climatological reality.[39] In this role, these experts function as important arbiters of ecopolitical conditions; they are people attuned to the processual relationship between human demands and desires and the life-spaces and other species that, at least from the purview of policy, are managed under the dominion of human choices about their future. If the impulse of conservation is to perfect the art of being affected, then through their professionalization, Saul and Alberto have become conservationally "attuned" to species.

There are multiple motives that animate the work of manifestaciones, bureaucrats, and experts in their defense of certain species and their environmental systems. Protections of certain species and particular spaces must live in parallel with Mexico's commitment to make renewable energy a central element of national policy. This, in turn, articulates with other, international policy missions to protect certain domains (such as atmosphere and climate) and other specific species (such as humans and particular kinds of flora and fauna). This is why, Alberto says, new technologies like wind parks and turbines were installed "without thinking much about their effect." Saul, conversely, underscores that *petroleo*-based production of electricity is "very dirty," and thus its attenuation is a priority for species protection. Even within the machinic protocols of manifestaciones and labyrinthine office hallways, distinct attunements to species and thoughtful calculations resting on an unnamed "greater good" can continue to exist.[40] It is, for Alberto, all about the "lens you use."

> I tell you, it depends on how you see it, or what kind of lens you use. Is it to the benefit or to the detriment of the environment? In the end you say, "Renewable energy, that is good for the planet." But let me give you an example of a hydroelectric plant. A hydroelectric project constructed on a river, they say this is environmentally friendly, but that is a lie. That is a lie because, in the end, if you have marshes downstream, if you have mangrove stands, what gives life to the mangroves and the marshes is the ecological flow [of the river], the hydrological runway. If, upstream, you are erecting an electric project, then you are altering the ecological flow of the water channel and an important downstream reserve of nature. Now you have no endemic species in the river. They are endangered with extinction; an already-fragile ecosystem is accelerating its disappearance. This affects, in different ways, the quality of the microfauna and the ecosystem. This begins to render a change in that system. Then, in that case, [renewable energy projects] are not supporting, or kind to, the environment. The impacts are, in fact, perhaps more drastic than that of climate change or global warming.
>
> That is why I say it depends on what lens you use. If you are flying the flag of climate change, then, yes, you are going to say, "Well, yes!"

But you have to analyze it systemically and comprehensively—each of these anthropogenic activities. So now we go to wind, and you say, "Okay it's kind to the environment, so let's do that." But if you put projects in the migratory route of the birds, or the route of the monarch butterfly passing through Mexico and the United States . . . then you wonder why they are not arriving during their migration season. . . . Why? Because they'll be stuck on their route, between wind farms in both the US and Mexico. They will delay too long in the middle of their passage and it will be an arduous return. When turbines are in areas where they need to descend, then you are putting a species in danger of extinction because their habitats are fragmented.

Alberto struggles with what appears to be a sacrificial offering: the local environment for the maintenance of a (dubious) global one. He appears aware that creatures and places, materials and flows, have been swept into a human course of unfolding events that are vague in their coordinates; as someone charged with protecting nonhuman others from (some) people's ambitions, Alberto is privy to the suspicions and doubts that are part of an Anthropocenic age. An attention to policy protections and the prevention of extinctions—which includes slowing the pace of global floral and faunal demise—is part of his conscious presentation of self. He, Saul, and others working in these spheres of care and management are deeply aware of their ambivalent roles, as they are caught up in the middle environmentally precarious times. Signs of species disappearance and the plodding on of power development stoke an awareness of creatural and habitational care and the associated horizons of risk. In the role of a conservation professional, and in the job of "environmental management," the rubric of care and concern has expanded and become more crowded: human needs, social needs, energy needs, development needs, the need for economic growth. The system has become very needy. And yet, people like Alberto are tasked with ensuring that ecosystems will continue to flow and to flourish.

When the scope of species awareness is extended, nonhuman beings start to feel increasingly wrapped up in our "world," perhaps especially for those who are being with species day in and day out.[41] New demands on the soils, waters, and skies appear through spreadsheets and carefully thumbed-through manuals, lists, and precautions. Each turn of the page indexes competing interests between creatures and plant life, energy and uncorrupted skies, local environmental stability and translocal measures to clear the air.

Birds—whether as predators or prey, appearing beautiful or hideous—are readily visible, "wild" creatures that exist as a quotidian part of most human lives. They are perhaps the most evident and familiar kind of wildlife in the world from a human's point of view. In one form or another, they occupy all habitats. Appearing everywhere, in spaces urban and rural, most birds are decidedly mobile. They are wildly in motion. Often, they are the only untamed animal in the most mundane contexts of human high modernity— like the pigeon in Times Square, fitfully pacing a line between the domesticated and the untamed. Birds also float through the most remote reaches of human dwelling—like the vulture circling a denuded peak in the high Andes. New World vultures are part of the family *Cathartidae*, derived from the Greek term for "purifier;" they are responsible for cleaning the dead and gorging on carrion. Human tendencies to associate vultures with death also serve as an uncanny metaphor for bird life writ large. Of the 9,865 known species of birds, approximately 12 percent are considered threatened with extinction.[42] Another 2 percent, or 192 species, are at an "extremely high risk" of extinction due to loss of habitat, toxic conditions, and for some, a changing climate. Indeed we humans know more about bird extinctions than we do any other creature.

Avian Risk

Biogeographically, the particular shape and orientation of the isthmus limits species distribution. These same qualities also make the isthmus an important migratory corridor for birds that utilize the region as a crossing point between the lowlands of the Atlantic and the Pacific.

The Isthmus of Tehuantepec is classified as an endemic bird area by Birdlife International, a global conservation partnership linking organizations to protect bird species and habitats. Endemic bird areas—overlapping breeding ranges of two or more land-bird species—number 218 in the world and are largely located in the tropics and subtropics.[43] In the skies above the site where the Mareña project was to be built, birds travel through a total of three migratory routes to move between North America and Mexico or farther south. An early "avian risk assessment" developed in 2004 by a US-based organization detailed the various forces that would threaten bird lives in the core wind zone of the isthmus if the development of wind parks were to continue. The assessment was guided by a mandate to create a framework to "preserve special species" and to document those species that fell under Mexican federal categories of risk: E = endangered (*en peligro*), T = threatened (*amenazadas*), or SC = special concern (*sujetas a protección especial*). A literature review was compiled and databases examined for information relevant to the environmental and biotic particularities of the isthmus. Ornithological researchers also visited the test site—1.5 miles west of La Ventosa—over the course of four days in mid-October 2004 and determined that the site held eighty-two species of land birds, raptors, and water birds. These are the names of some.

white-winged dove
orange-breasted bunting
white-lored gnatcatcher
plain-breasted ground-dove
common ground-dove
west Mexican chachalaca
stripe-headed sparrow
cinnamon-tailed sparrow
turkey vulture
black vulture
roadside hawk
crested caracara
hook-billed kite
gray hawk
short-tailed hawk
white-tailed hawk
red-tailed hawk
laughing falcon
crane hawk
common black hawk
great black hawk
solitary eagle
peregrine falcon
Harris's hawk
collared forest-falcon
osprey
Mississippi kite
sharp-shinned hawk

Cooper's hawk
American kestrel
merlin
bat falcon

Because of its inland location, the site for the proposed wind park that the ornithologists surveyed did not have a large water-bird population, although sixty American white pelicans, 150 Wood Storks (designated as "special concern"), and three hundred Franklin's gulls were spotted in migration. The cinnamon-tailed sparrow—also known as Sumicrast's sparrow—was one of the most common permanent residents of the site and is classified as "endangered" in Mexico.[44] This was judged to be "noteworthy"; further nesting studies would need to be carried out.

The resulting reports claimed that the cinnamon-tailed sparrow population in the isthmus was generally robust. Experts were thus not much troubled by habitat loss, which was again deemed to be "unlikely to constitute a biologically significant" impact. However, SEMARNAT requested additional baseline surveys to document flight patterns of other bird species over the period of one year in Santa María, revealing yet more species:

northern bobwhite (*Colinus virginianus*) (near threatened)
elegant tern (*Talasseus elegans*) (near threatened)
reddish egret (*Egretta rufescens*) (near threatened)
painted bunting (*Passerina ciris*) (special concern)

Whichever lens one looked through, more species seemed to be appearing, more bodies in motion across the sky and sea. The lists would begin to read like a requiem in the future subjunctive. Budding extinction narratives such as these can easily take the form of elegiac repetition as they reify the names or the nomenclatures of those deceased.[45] Grand lists of species also perform a bureaucratic mode of encounter: thinking with species as multiple categories that can be assessed and evaluated on the flat plane of paper and ink. In lists such as these we find the union of diagnostic capacities (threats or concerns), elegiac mourning (for birds not yet dead), and an accounting (that numericizes birds' status and situates them in the scale of value).

Species exist across domains of empirical observation, descriptive note taking, protocol adherence, and the ability to be located in a particular place and time. In the isthmus, birds must be sighted (seen) by humans as well as sited (located) in, on, or over the landmasses in question and in the documents that matter. The 2004 assessment notes, for example, that during migratory periods, tens of thousands of birds are present over the project site. Although migrating raptors normally fly at relatively high altitudes, the research visits indicated that raptors were flying at lower altitudes through their migration corridor—between 100 and 490 feet. Their lower-than-average flight pattern was said to be a response to the strong winds on the Pacific side of the isthmus. Water birds, like raptors, were also found to be flying at lower elevations over the project site, evading the powerful winds. Hundreds of migrating raptors were seen settling in the forest at the project site overnight, and it was becoming clear that the Mareña park would be in the middle of a raptor migration corridor. Because strong winds worked as a pressure tunnel, forcing birds to fly at lower altitudes, the species transiting this airspace would be at increased risk for collision with turbine blades. For birds of prey that frequented the area for hunting, seeking sustenance would also be increasingly perilous.

Both the 2004 avian risk assessment and the one prepared for the Mareña location were, however, reluctant to declare extensive potential impact to bird populations. Each report deemed that there were no "demonstrated biologically significant impacts" either on a case-by-case basis (for each site) or cumulatively. Data from wind power projects in Europe and the United States was then used in a corollary way to make recommendations about the Mareña site. This data may not have been "cut and pasted" as rumors had suggested, but using evidence from other wind parks, thousands of miles away and bearing different biogeographic conditions, would seem to be an equally illegitimate practice. The final verdict was that it was "likely" that little harm would befall isthmus bird populations, predicated in part on studies conducted in entirely different parts of the world. Qualifications and uncertainties also riddled the pages of the impact assessments. The 2004 report stated, for example, "Given the lack of studies, the extent of nesting displacement is hard to predict," and, "Significant questions are how flight strategies vary in different wind conditions and how migrating hawks use the site during spring migration." Birds' spring migration patterns themselves were described as "unknown," and nocturnal migration studies in the region were nonexistent. Birds were certainly quantified and identified throughout the studies, but ultimately birds in the wind zone were less an object of conservation than an object of experimentation.

It was determined, ultimately, that bird mortalities caused by isthmus turbines would render valuable scientific data. "Collision fatalities" that would

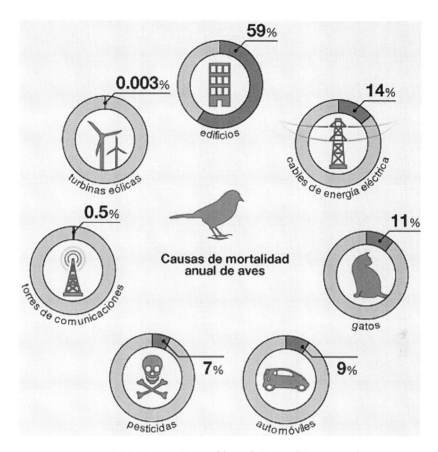

59%
edificios

0.003%
turbinas eólicas

14%
cables de energía eléctrica

0.5%
torres de comunicaciones

Causas de mortalidad
anual de aves

11%
gatos

7%
pesticidas

9%
automóviles

FIGURE 5.3. Mareña Renovables website graphic: causes of
annual bird mortalities in the isthmus

be documented once the park was in operation would, in a morbid twist of environmental authority, offer useful statistical information to be applied elsewhere. "Monitoring results of birds' mortalities" in two projects currently under supervision by the Interamerican Development Bank, the report noted, when "combined with the results of this Project will be helpful in determining the extent of cumulative impacts." Killed birds would provide, in other words, "a baseline" for future wind park developments across Mexico and Central America.

Isthmus wind parks could thus become an experimental lab, a site of animal testing with turbines; isthmus bird fatalities would serve as critical baselines. The Oaxacan foray into superdense wind park development is, in the first instance, based on a desire for cleaner power sources. But it is equally a matter of weighing local deaths against global climatological risk reduction. Experimental energy forms produce a macabre twin when they also function as an exercise in animal experimentation.

Among all the enunciations of bird-related risk, however, none were as bizarre as a graphic that was posted on the Mareña website. The illustration depicts cats and birds, turbines and buildings, wires and towers, cars and pesticides, each category quantified by its due percentage of bird-kill rate. Turbines, the graphic shows, pose nearly no danger to birds; a mere .003 percent of bird fatalities could be attributed to the turning of blades. Where buildings are the greatest danger overall, killer cats have a venerable place in the metrics as well, though it tests methodological credulity to imagine the study of assassin cats and their feathered victims that so neatly equated 11 percent.

Blind Mice

Birds are not the exclusive inhabitants of skyspaces, even if they are most readily seen by human eyes. Birds' nocturnal, mammalian analogues—the creatures that English speakers call bats—are in Spanish called *murciélagos*. The etymology traces back to the Latin *mus*, or "mouse," and *caecus*, "blind." Like birds, bats are a planetary ubiquity, occupying nearly every ecosystem on earth and comprising one-fifth of all terrestrial mammals.[46] Bat populations are particularly rich in the southern Mexican isthmus, Central America, and the tropically forested Southern Cone. And while most humans do not have the proximate, daily encounter with bats that they do with birds, they are everywhere around us at all times.

Adriana Aragon Tapia, the subdirector of renewable energies at SEMARNAT, considers murciélagos to be a "resource of the wind" and a Oaxacan "treasure" in need of protection. The terms "treasure" and "resource" reflect the ecosystemic function of bats in the isthmus and elsewhere, or everywhere.[47] They diminish insect populations that may harbor diseases that threaten humans and other forms of life. They provide especially nutritious waste for plants, and they are hardy disseminators of seed stock, providing an essential reproductive labor in the lives of trees such as those belonging to the genus *Cecropia*, which are called *guarumo* in Spanish. In fact, seed dispersion by bats was found to surpass that of birds in forests near the isthmus.[48] Bats, like birds, are also prone to certain kinds of perils. While they may be less likely to be struck from the sky by turbine blades, bats can have the life blown out of them.[49]

Barotrauma is a fatal fate for bats swerving through corridors of turbines. Many of the dead bats that litter the grounds of wind farms have never suffered external physical trauma, but inside, their lungs have been exploded. Unlike birds that have rigid lungs, bat lungs are pliable. When these more fragile lungs are exposed to a sudden change of atmospheric pressure—such as that occurring directly in front of a rotor in operation—a bat's lungs will expand quickly beyond their capacity, and the creature will drop dead from the sky. As if their susceptibility to barotrauma were not dangerous enough, bats are also, in fact, drawn to the blades of turbines.

Various hypotheses have been offered as to why bats would be attracted to wind turbines. In Canada and the United States, thermal imaging has documented bats attempting to land on turbines, which they may perceive as roosting trees. The structures may also be taken as a source of food since the blades and rotor area are peppered with dead insects. Another source of attraction for bats may be the heat generated by turbines. Or it may be that sound frequencies and electromagnetic waves produced by turbines disrupt echolocation, causing bats to inadvertently hurtle toward towers and blades. According to comparative data from wind parks in North America, Europe, and other Interamerican Development Bank–financed parks in Oaxaca, bat mortalities outnumber those of birds. Despite these comparative reports, the IDB assessment nevertheless concluded—in a final morbid deferral— that "there is no way to know" whether bat populations would be affected by the installation of a massive wind park such as that proposed by Mareña.

While the likely kill rate of bats may have been diagnosed as "unknowable," there was an attempt to at least ascertain what sorts of bats would potentially be put at risk. Distinguishing species was a key element in this exercise. Over the course of twelve months, using acoustic detection, infrared observation, and net capture techniques, several bat species were identified across the barra and in the sites where the towers were to be placed. The lesser long-nosed bat (*Leptonycteris yerbabuenae*), a bat considered "vulnerable" by the IUCN, was spotted repeatedly at the San Dionisio site. The lesser long-nosed bat had experienced a 30 percent population decline over the last decade due to a shrinking habitat of thorn scrub and deciduous forest. Over nine nights of acoustic detection—three in summer, three in autumn, and three in winter—a total of 184 bat passages were recorded in the area where Mareña's turbines were planned. Northern yellow bats (*Lasiurus intermedius*) flew through 144 times, Pallas's mastiff bat (*Molossus molossus*)

FIGURE 5.4.
Leptonycteris verbabue-nae, the lesser long-nosed bat. Illustration by Mario Norton.

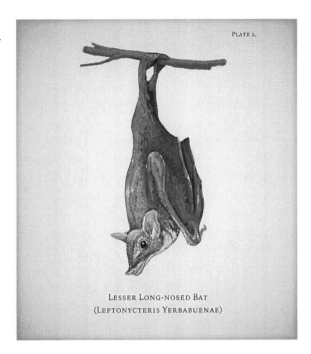

PLATE 2.

LESSER LONG-NOSED BAT
(LEPTONYCTERIS YERBABUENAE)

made thirty passages, the black mastiff bat (*Molossus rufus*) made eight, and the mouse-eared bat (*Myotis*) a mere two. The lesser long-nosed bat, a migratory species, was the most abundant of all.

The lesser long-nosed bat travels down the Pacific Coast, following a nectar route overlapping the proposed wind park footprint. Seeking out agave and other forms of nutritive sustenance, the lesser long-nosed was at risk. But instead of a solution, the report offered an experimental proposition. "Monitoring during construction and post-construction will help to determine the precise habitat of the Lesser Long-nosed bat," it noted, and this would, in turn, allow for observation of the "extent of barotraumas." Dead bats would, in other words, at least provide good data. Bats' biotechnological skills of echolocation may be impossibly challenged by energy infrastructures erected in their paths of flight. Their perishing, in turn, makes it possible for human calculations of their species' plight.

Birds have long been understood to have predictive virtues and, at times, the ability to illustrate futures. In many indigenous communities, birds are said to sketch pictures in the sky, moved in one arc or another by the spirit of the place over which they fly.[50] Roman divinations were also drawn from the observations of bird flight (auspices) and were integral to foundational legends. No important judgments were made without consulting the augur. Likewise, Roman oracles were also consulted for their omen, or "true speech," a truth predicated not on the present but on an emerging future. Epidemics, disasters, earthquakes, and monstrous births were understood as prodigal signs sent by the gods to convey displeasure; birds and men could read these futures.[51] Weaving across the sky in the same way they did centuries ago, species of birds and bats in the renewable age seem to offer similar revelations and prognostications. They are unwilling indicators of harm and risk. Bird and bat bodies accrue on one end of a scale that is balanced against the ephemeral good of wind power. Certain creatural deaths get caught up in the rotations of power moving forward, and lives are quantified, and thus qualified, on a scale of importance. Varieties of sky life, like bats with their barotrauma and birds on the wing, also function as indicator species with a morose experimental interest. The edges and limits of their vulnerability to "significant biological" harm become summoned as baselines and quantifications of the necro-possible: the prospect of extinction in the future subjunctive. Dead bats and birds are thus enabled, as species and as anticipatory subjects, to be evaluated against atmospheric goals for a "greater good."

If airborne creatures are vivid in the charts, graphs, and rhetoric of species in times of transition, the aquaspheres of lagoons and seas are equally dense with species, both those that are valued and known as well as those whose existence is obscure, ebbing into the forgotten.

Water's Edge

Fecal matter, human and nonhuman, is simply a part of the hydrosphere in the isthmus. The watery spaces of rivers, lagoons, and creeks across the region have absorbed the effects of constant human contact and contamination; cattle ranching, agriculture, and consumption have also added themselves to the current chemical composition of land, water, and air.[52] As the IDB report declared, the isthmus has "been exposed to intense human activities" over the past several decades. This has led to a deterioration of "the 'naturalness' of the area." While lagoonal and sea waters were never designated for turbine placement in the Mareña plan, it was water and the lives within it that were repeatedly invoked in debates about wind power mechanisms and their electric offspring. The interrelationship between water, wind, and species reveals not only the habitational importance of hydrospaces but also the rhetorical power of particular species. For, in the end, water-dwelling species, more than windborne or land-bound varieties, became repositories for the most sustained conflicts between humans and their species-others in the battles over the Mareña Renovables project.

The National Biodiversity Commission of Mexico (CONABIO) has designated the isthmus lagoon system as a "priority marine region." Mareña's "direct area of influence" includes ikojts indigenous communities in San Dionisio del Mar, San Dionisio Pueblo Viejo, and Santa María del Mar, and all appear in the environmental impact statement. Among these communities, the report finds, "the main activity is fishing." In Santa María del Mar, the production of "shell based handicrafts" is listed as a primary pursuit, along with subsistence agriculture and, again, fishing. Environmental attention is given to these practices in name, and yet a strange aporia ruled the logic of the environmental assessment. Although humans are listed among those species "impacted," there was no attention given to the fish, shrimp, or other shelled creatures of the lagoons and sea on which these particular humans depend. At no point is there a consideration of fish in the manifestación. Land creatures (such as jackrabbits and iguanas) and skyborne creatures (birds and bats) all make it into the pages of these reports. But while "fishing" is indicated as a common livelihood, "fish" are not mentioned. This was a collective failure of conservationist bureaus and international developers alike: the inability to see through murky waters and respond to the nexus of fish, shrimp, and humans. And it proved to be a fatal oversight due to which the Mareña project met its watery end.

It is not that water in itself was ignored as a potentially threatened habitat in wind power development. It was merely a question of what kinds of species were being sought in these watery spaces. Marine turtles, for one, were christened as a "species of conservation importance." Unlike fish, they would be given what was described as "special attention" in environmental impact reporting and protections. Where the park was to be sited is also the place where several species of marine turtles make their nests and hatch their eggs. Sandy dunes, near to the sloping edge of the barra, create a natural infrastructure for these rather-massive floating and paddling creatures. Both the Laguna Inferior and the Gulf of Tehuantepec constitute an integral home range for three species of marine turtles: the black turtle (or the Galapagos green turtle, *Chelonia agassizi*), considered endangered; the leatherback (*Dermochelys coriacea*), designated as critically endangered; and the olive ridley turtle (*Lepidochelys olivacea*), classified as vulnerable by the IUCN. Mexico's seven sea turtle species are protected by the Convention on International Trade in Endangered Species of Wild Fauna and Flora.

SEMARNAT requested sea turtle population studies in order to ascertain which and how many turtles were near the wind park sites. Using indirect observations—such as tracks on the sand and the presence of nesting areas—it was confirmed that both olive and ridley populations had been on land on the southern side of the village of Santa María. The leatherback was unaccounted for. Worries about turtles were several and compound. Nesting grounds needed to be preserved and undisturbed during crucial laying and hatching seasons. Another concern was the maritime traffic that would increase during the construction phase of the Mareña park, which would increase the risk of turtles colliding with barges (carrying turbine mechanisms, towers, and blades) and boats (shuttling laborers or other materials). Due to their precarious status and the dangers enumerated by several studies, turtles became unique objects of attention. But it was also the case that turtles were already established subjects of environmental concern and conflict. Sea turtles are commonly defined as "charismatic" species: ones that humans have a particular affection for and feel an obligation toward.[53] Sea turtles had achieved a particular form of "biolegitimacy," falling within the lines of a life that is seen as eligible for exceptional protection.[54] Though they garnered human conservation attention, turtles were also caught up in the nets of economic and social life in the isthmus.

Ping-Pong

Outdoor food markets are part of each day in the center of Juchitán. There, available to be touched, tested, and sometimes sampled are *nancites*, *bananos*, *uvas*, *fresas*, *chayotes*, *nopales*, *tomates*, and meats of every kind and cut, from hooves to snouts. And there are also always eggs, most from chickens, but many, many from sea turtles. By some accounts, turtle eggs are believed to have aphrodisiac powers, and they are sold for about a dollar each. Round metal pans brimming with what look like slightly dented Ping-Pong balls are a central feature of the Juchitán market at the right time of year, and the vendors, almost all of them women, will entreat you to just try one for yourself to experience this local delicacy.[55] Harvesting marine turtle eggs is officially illegal in Mexico, though it is rarely prosecuted. In at least one case, however, it was. As I sat with Magda one afternoon aboard a bus bound for Mexico City, she told me how her sister's fate had been bound up with turtle eggs. Her sister, who, like Magda, was staunchly opposed to the Mareña project, had been denounced by rivals in the community who, Magda made clear, were also keen supporters of the wind park's development. Both sisters were convinced that the prosecution that followed had little to do with turtle eggs and everything to do with wind power politics. Magda's sister was well-known in the community as a purveyor of well-cured turtle eggs that she gathered by moonlight from nests on the beach. Apprehended by the police one night while doing so, she was promptly jailed, and there she remained. Magda made very clear that she saw this as a deep injustice; her sister had been martyred for the cause of turtle eggs—eggs that were suspiciously linked, for each of the women, to the wind park. For Magda, turtle egg collecting was an occasional, seasonal economic opportunity, and thus her sister's imprisonment seemed an unnecessarily cruel outcome and one that she saw as unequivocally tied to the politics of wind power.

The human threat to not-yet-hatched turtles and to those that are poached as adults is not a new phenomenon in Mexico. The environmental impact statement for the Mareña project noted that the primary hazard for turtles in the site area are local communities that collect turtle eggs for human consumption. In response to the risks posed by turtle egg gathering, the company declared a promise to SEMARNAT in the interest of moving its project forward. In an effort to protect turtle eggs, Mareña announced they would finance patrols of the sandy edges of land where the project would be located. Every two hours, security officers would tramp the beaches. All night long, from eight at night until eight in the morning, Mareña security teams would be tasked with halting egg poachers. In other words, turtle egg hunters would be hunted by company guards. This was a possibility in the future subjunctive that did not sit well with many isthmus residents.

Environmental Degradation and Its Authorities

Environmental impact assessments appear to be objective analyses, determined through the empirical arts. More precisely, however, environmental impact is instead a subjective diagnosis that raises questions about how "environments," nature, and harm are perceived and calculated. For those protesting the further development of wind parks, the potential for increased environmental degradation by the turbines was a clear danger. But the form of ecospheric damage was of a different quality and kind than that imagined by SEMARNAT, developers, or environmental diagnosticians. Where the demands of state policy and environmental reporting centered on various indexical species (hares, birds, bats, and turtles), proclamations issued by those protesting wind parks focused attention on the interrelatedness of ecosystems and their human inhabitants. For many of those opposed to furthering the development of wind power in the region, concerns rested less on singular species than upon the ways that a handful of species—namely, humans and waterborne life—were enmeshed with one another in ecosystemic terms. One missive, titled the "Public Denouncement by San Dionisio del Mar," made this position quite clear.

> A commission convened by the people's general assembly made a reconnaissance tour of the barra and we noted several traces of the company [Mareña] on our territory. Said company also continues harassing us, and threatening to enter the barra by any means necessary. To this we again state that the ikojts people of San Dionisio do not want this project because it is a project that trespasses on our territory and robs us of our land. Once it is installed, the only monitoring to take place in the area will be in service of protecting the monetary investments of the company. They want to place on our island, at the heart of the upper lagoon of the Isthmus of Tehuantepec, 102 wind turbines, 2 electric substations and five docks that will involve 266 boat trips a day. These boat trips will involve the use of diesel and will profoundly affect the marine ecosystem, which is considered by CONABIO to be a priority region for its high marine biodiversity.

The resistance in San Dionisio del Mar saw risks to marine ecosystems that closely paralleled the theft and occupation of their land. On the other side of the debate, one of Mareña's employees, Eda, claimed that these accounts of potential environmental wreckage were disingenuous and suspect. Her perspective, which echoed that of others positively inclined toward the Mareña project, centered attention on the profound environmental damage that already existed across the lagoonal area. More important still for Eda and others, was that the claim to environmental injustice and assertions of conservationist principles by the resistance were baseless. In a nostalgic turn, she called upon her own memories of isthmus waters, its creatures, and its humans. "I will tell you," she began, "the laguna is so polluted now anyway." She explained:

> It is not like it was when I was a girl and we went to Playa Vicente and La Punta del Agua. Those were our beaches, and my father used to have to carry us out past the mussel and abalone shells on the shoreline so that

we could play in the water. Now there are five polluted rivers drain-
ing into the lagoon. It is full of human waste, and it stinks. And we
import a lot of fish here now. Whole species that we used to eat from
the lagoon can't survive there anymore because of the contamination.

With a more pristine past in mind, Eda went on to attack what she believed
were false claims of environmental concern. "Why," she implored, "are the
antieólic activists so malicious? If they are opposed to private investment,
just say that. Don't invent all these lies about environmental effects to tell
people." She went on, "At most, the barges will make six trips a day [during
the construction phase], not the 266 trips a day the resistance claims."

"Why do they lie so much?" she wondered aloud several times.

It was also important, Eda explained, that we know that the company
would be planting red mangroves, one of the indicator species of the Anthropocene that has declined by half globally because of human effects
on coastal wetlands.[56] Eda spoke about another small tree that locals had
been harvesting for decades in order to cook or smoke fish. This tradition
she believed, was more destructive to the species than any wind park would
be. Eda was full of environmental worries about mangroves and human
contamination, and she was deeply skeptical of protestors whom she believed were shrouding their opposition to private investment in the guise of
environmentalism.

Appropriating an ecological rationale, however, was a wise step for those
resisting wind developments; in fact, it followed the logic of SEMARNAT to
the letter. Because the agency would "not take complaints in the form of 'I
don't want this [wind park] because I don't want it,'" critiques needed to be
framed through an ecosystemic rhetoric; they needed to indicate environmental knowledge and authority. Where Eda gestures to species of trees,
denouncements by the opposition emphasize dangers to lagoonal waters
that humans require in order to continue to fish and gather shrimp. Beatriz Gutierrez Luis, a teacher from San Mateo and an opponent of the wind
parks, put it this way: "I understand this is supposed to be a form of clean
energy. [But] if they gave us all the money in the world, we'd still say no. Our
children and our grandchildren will depend on the fish, the shrimp, the love
of the land, respect for nature, and all of our cosmology that we have as an
indigenous community."

A connection between nature and indigenous human occupants took
shape as a narrative that was repeated many, many times in different ways
and forms in the isthmus. An emphasis on the cosmological union between

humans and the ecological spaces in which they interact came again in a statement issued by the San Dionisian resistance. "We have a spiritual relationship with our lands, territories, seas, and natural resources, which form the collective property of our people and our community," they wrote. "Therefore, our territory is not a commodity that can be sold, rented, or privatized."

Forging a link between indigenous peoples and natural environments is not a new claim; neither is the pronouncement that many indigenous communities hold a unique environmental sensitivity that refuses monetization. Popular culture and the anthropological record are full of associations between First Nations peoples and a sometimes abstract, sometimes very tangible "nature." Some of these interpretations are predicated on sound evidence and actual practices, and some of them are drafted from stereotypes alone. However, the set of associations that map indigenous people as one with the land, sky, and sea are only one instantiation of eco-human relationality. What also surfaces among both those voicing opposition to and support for the wind park is a diagnostic value scale that takes "the environment" as an object of management. For the bureaucrats at SEMARNAT, as for the indigenous fisherfolk of San Dionisio, the environment is a locus of care and protection. Species figure heavily across these discussions and documents. In this context of environmentally oriented claims, certain kinds of life forms are attributed status and importance as species. For wind park developers, these are sea turtles and mangroves. For local fisherfolk and indigenous residents, they are humans and fish. Defending one species against another—whether in the name of protecting human practices or mangrove habitats—is a way of being with species: enunciating their precarity as either living entities or living practices.

Ending with Species

───────

To build a wind park, one ought to begin with species. This has not always been the case, and it has not been the norm for many human expansions and infrastructural projects that cement over life-spaces. But we species occupy a different niche now. Species sentience, or a creatural awareness, feels as though it inhabits each move and practice in environmentally precarious times. Being with species—jackrabbits and raptors, turtles and bats, fish and humans—highlights how particular value scales become created. Taxonomies and categories of worth are added or subtracted for certain kinds of

Sharp-shinned Hawk.
1. Male; 2. Female.

FIGURE 5.5. *Accipiter striatus*, sharp-shinned hawk

beings, both human and nonhuman. To be sure, placing species in a relational cast of worth is not a new human act. In fact, one could argue that this has been *the* foundational practice of settler colonialism. Some humans have been privileging certain nonhuman species over others for millennia. And some humans have been privileging certain humans over others for millennia as well. Species' ranks may be distinct in different times and places, but what appears to be universal is that most humans are prone to set themselves at the apex of that scale, even in those societies where mutuality between humans and nonhumans is especially respected.[57] Changing earthly conditions, however, demand different ways of being with species.

Whether the sharp-shinned hawk is killed by a turbine blade or befalls some other fate should not matter any more or less now than it did when humans first trapped wind to power machines in the first century AD. But I would argue that it does. In a climate-imperiled time when humans anxiously encounter their own environmental survival, we live in a different world. Between human articulations of displacement and the bioscientific management of species through "conservation importance," we find ethnographic parallels between humans and their other-than-human analogues. Every one is concerned, and everyone is an object of concern, whether we are *we* or the bat herself or the sparrow himself. Beyond merely thinking within the classificatory category of "species," we might open to another possibility, a possibility of mutuality that comes, at least in part, through shared fears of extinction. Conditions like these challenge us to account for how we ought to approach the evaluation of species, including our own. Human politics may have seeped into every being's and every thing's future, but a recursive truth also exists: the fate of extinction that nonhumans have historically suffered from human acts now threatens to be human destiny as well. What used to be animal problems are now human problems. Being with species is an awareness of these kinds of mutualities. We are all species now.

If wind power has been part and parcel of forces that have driven toward the Anthropocene, and carbon-fueled machines occupy the zenith of accelerationist conceits, then it is surely species at the end of it all. That recognition binds us to the fact that modernity tried to forget: we are contoured by reciprocal complexity all the way down. Like turtles. Or turtle eggs, delicate and papery under a scorching sun.

6. Wind Power, in Suspension

Por tradición somos luchadores. —UNIDENTIFIED BINNIZÁ MAN,
SANTA MARÍA XADANI, OAXACA

The Treasure of the Isthmus

The Mareña Renovables website was a deep digital pool filled with ecologically authoritative proclamations. It conveyed great ambitions for the wind of the isthmus, heralding it as a *"tesoro"* or "treasure" that would "bring Mexico into the future of sustainable electric energy generation." With a carefully crafted set of attentions, the company's website drew lines of redemption from the Barra de Santa Teresa and its waters to the greater planetary biosphere. In this space, the isthmus wind would offer itself to remedy the climate and save a little part of the world. Balancing upon neoliberal development priorities as well as biopolitical works, the wind park was touted to be renewable in two ways: offering sustainable power and enhancing "the well-being of the communities in the Isthmus of Tehuantepec." Guarantees of local development coupled with greenhouse gas reductions seemed wholly laudable, and the company appeared to have moral authority on its side, founded on the goal of improving the future of a shared climate.

Fisherfolk from around the region, however, were concerned less with greenhouse gasses than they were with troubled waters. Ruined waters would come with the turbines, they feared, through light, noise, vibration, and mud. Sludge stirred up during the construction phase of the park might result in

FIGURE 6.1. Computer-generated depiction of what the Mareña
park might have looked like

meager fish and shrimp harvests, and lights atop the turbines would shine
upon the lagoonal waters and potentially scare away fish.[1] Noise emanating
from the giant machines, it was thought, would also frighten the fish. And
the vibration created by the gesticulations of turbine blades would surely
shake the barra, the silty floor of the laguna, and the sandy ground of the
Pacific, causing fish and shrimp to move to other watery places farther away
and perhaps out of fishermen's range altogether.

In this, the final chapter in the story of Mareña Renovables, the park
hangs in suspension. Supporters of the park had come to see opposition to it
as motivated by political opportunism, specious environmental claims, and
meddling "outsiders." For those resisting the park's construction, Mareña's
backers were inspired only by their profit margins; they were ruthless interlopers bent on extracting wealth from the istmo wind but rarely concerned
with the impacts the park would have on local people and ecologies. In the
making and disassembling of the giant wind park, failures of attunement
abounded. Histories of insurrection and displacement were never given
their due, and perhaps more importantly, the imagined futures of local residents went unrecognized for far too long. The wind park's fate was tied to

a series of political maneuvers by caciques and corporate representatives, government officials and indigenous activists. But in the end, wind power would collapse into the waters that sustained both people and fish.

FISHERFOLKS' APPREHENSIONS ABOUT the aquaspheres on which many depended were taken up loudly and repeatedly in the voices of the resistance. Wives and families of fishermen also pronounced their uncertainties about how fish would fare with a dramatically transformed barra, now industrialized with electric infrastructure. Many women in Juchitán, for instance, earn their living through the fish trade: drying, slicing, and selling the day's catch. Every fisherman with whom we spoke—ikojts, binnizá, and mestizo alike—shared palpable apprehension about noise and vibration, light and mud. The construction phase of the park would be one disruption, but their concerns were attuned to the more enduring effects of the turbines over the expected three decades that they would be in operation.

Across the isthmus, fishermen also confided to us that fish stock had been on the decline, in part due to pollution and in part due to overfishing. Placing massive turbines atop the narrow sandbar of Santa Teresa, they believed, would further endanger their ability to fish both for subsistence and for a modest income. Fisherfolks' concerns centered on the potential environmental damage that the wind park might bring. Their worries about the environment were also directly tied to their livelihoods and their capacity to work with, in, and around the waters that surrounded their homes. More often than not, the welfare and well-being of future generations was present across the stories people told of uncertain futures.

The unique geographic location for the proposed Mareña park made its potential environmental impacts difficult to fully estimate. It would have been the only wind park in the world that would occupy a sandbar. This singularity, along with doubts about the efficacy and honesty of governmental agencies and political actors, undercut the value of environmental impact reports and permits; the reports' transparency was questionable and their findings, for many, were equally dubious. While the Interamerican Development Bank's Environmental and Social Management Report acknowledged the possibility of short-term "economic displacement" for fisherfolk during the construction phase of the park, it did not analyze the long-term impacts of the park's presence on local populations, both human and fish.

An absence of scientific data about the conditions specific to the Mareña site, coupled with misgivings about the ability to accurately diagnose the damage that might occur, meant that uncertainties multiplied. Manuel, speaking with us in San Dionisio, summarized these worries well: "The wealth of our sea, of our people, of our source of work and nourishment, is vital. If the wind project comes in, we will be buying foreign products coming from other places, which will make feeding ourselves more expensive." Manuel's understanding about the "wealth of our sea" turned out to be portentous. Over time, we began to increasingly hear the expression "the sea is our bank." This might have been a clever spin on the evident presence (or imposition) of banking interests and multinational capital that backed the Mareña project. But "the sea is our bank" was also an empirical statement. According to reports, there are approximately five thousand indigenous families that rely on fishing for their existence in the barra region. Even if the number of fisherfolk who survive by fishing alone are few in number, many residents around the water's edge depend on fishing in conditions of economic and food insecurity: if all else fails, if there is no work to be had, if food is short, the sea is always there, and you and your family can eat.

Mareña's management made an effort to forestall worries about aquaspheric ruin by offering onetime payments to fisherman for their lost income during the construction phase of the project. Newer and better boats and outboard motors were also offered as enticement. But the deal came as too little too late.

As voices of protest grew louder in late November 2012, Mareña developed a special page on its website dedicated to fishing cooperatives. "The conclusion [of] studies and international experience," it stated, "is that there are no effects on fishing caused by the operation of a wind park." The text went on to note that the construction phase would entail a great deal of movement and added turbulence in the water, but that the company had sought to make the construction process as rapid as possible to avoid any unnecessary disruptions. The sentence centered at the foot of the page in bold type conveyed the key message on offer: "The wind park project respects the culture of fishermen." A neat diagram assured all viewers that light interference would be minimal, noise would be equivalent to less than the hum of a refrigerator, and vibration would be absorbed by the cement foundations into which the turbines would be embedded.

Interestingly, no fisherman we encountered had ever actually seen the Mareña website; in fact, none knew it even existed. This was an irony that begged the question, Which audiences did the website hope to influence: curious reporters, international observers, or perhaps nervous investors?

FIGURE 6.2. Fishing skiff, Barra de Santa Teresa

Fishermen's concerns and their ecologically oriented statements were also not without their own environmental impact. Their trepidations about the continued viability of the region's waters were derived from an awareness of its aquaspheric limits; fishermen know the water and its inhabitants. Like the environmental impact report devised in Mexico City, the fishermen had their own form of environmental assessment: How much could an anthropogenically injured lagoon and sea be expected to yield, and would these waters continue to provide a minimal quotient of subsistence or income for local fisherfolk? As reasoned evaluations made by those who live by, for, and from the sea, these were legitimate doubts about the potential environmental degradation that would follow the park's installation. In a context of somewhat-sketchy scientific reporting, especially where no comparably sited park could be found, the environmental assertions of fishermen gained support among many in the region. This was not so everywhere, however.

Wind industry professionals, bankers, and government officials in the state and national capitals were generally quick to scoff at fishermen's claims regarding fish and shrimp. Their interpretations of the park's potential impact were most often dismissed as ignorant, superstitious, or at best, a form of indigenous knowledge that, while quaint, had no place within rational debate. Whether or not noise, vibration, light, and mud would have resulted in the deleterious outcomes that fishermen predicted remains unknown. But these uncertainties do uncover two important contingencies. They index the difficulty of producing convincing ecological knowledge when government and corporate interests appear to be compromised by financial gains rather than attentive to local environmental protection or sovereign rights. And perhaps more importantly, they index which knowledges are valued and by whom in a continuum of expertise and experience within ecologically precarious places.

Expiration

When Judge Coronado placed an injunction on the Mareña project in his humble office in Salina Cruz in early December 2012, he legally stalled the park's construction. But his declaration was, in fact, the beginning of the end. Although his order was only a provisional mechanism, a stay, it ultimately functioned as a death knell.

It would be another year of tense confrontations, death threats, and desperate attempts by the company and the state to rechannel the future of the park. But the disassembling had begun. National and local press outlets widely publicized the judge's decree, and speculations about the fate of wind power in the isthmus showed a new measure of uncertainty. Statements issued by Mareña representatives took on a decidedly panicked and, at the same time, accusatory tone. In the wake of the request for the amparo and injunction, opposition to the park was now being seen as an assault: against development, against the people of the isthmus, and perhaps most dramatically, against the future.

Accusations of blackmail (*chantaje*) against the resistance were now more blatantly emerging, the claim being that defiance against the park was simply a ruse to squeeze more money or political favors from the company and the state. While discourse of this kind had been circulating prior to the amparo and injunction, new denunciations of blackmail were pitched as a risk to the future. If state officials and others—the Mexican wind industry, for instance—were to allow opposition to continue, it would most likely spell the end of wind power in the isthmus and possibly in all of Mexico. As Edith put it, "This amparo is blackmail, and if this company leaves, others will definitely be reluctant to build here. The doubts and lack of confidence will multiply." She was convinced that investment in the istmo would surely collapse. To fail to support the big wind park, it was suggested, was to deny Oaxaca the potential of its treasure.

While members of the resistance were celebrating the judge's decision, commentary increased through the channels of the press. A handful of days after the amparo, the Oaxacan state began to weigh in more explicitly and

FIGURE 6.3. Cell phone message: "patria libre o morir"

forcefully, if diplomatically. The governor, Gabino Cué, warned that the amparo and injunction were sending a "bad signal," terrifying both current and future investors as well as dissuading them from financing projects in the state. "[Mareña's] investment represents one of the primary levers to achieve peace and progress," he explained, "and this is valued by the people of Oaxaca."[2] It was not only foreign capital that was being threatened, he averred, but harmony and advancement. While disparaging both the physical and legal blockades that were threatening the project, Cué also held the company accountable to its promises of development and investment, encouraging them to stay the course and "do right" by its stated obligations. The governor's plea, coupled with the legal blockade provided by the injunction, had an effect. By the third week of December, Mareña committed to pay three million pesos for a school in Álvaro Obregón. If opposition was, in fact, driven by chantaje, it was working. Biopolitical works were being offered afresh.

By the end of December, with a brief interlude to celebrate the Christmas holiday, the battle continued with a sham asamblea meeting in San Dionisio del Mar. The meeting, according to the wind park supporters who organized it, was an effort to establish a majority vote that would seek to get the injunction lifted. Instead of amicably voting, attendees at the meeting turned violent, and thirteen people suffered injuries when fistfights broke out and chairs were thrown. Supporters of the Mareña project and those in the opposition were becoming increasingly ferocious in their encounters with each other, even as the injunction had sought a legal, peaceful solution.

An Overdrawn State

Oaxacan government officials from the governor down were being pulled deeper into the trials of the Mareña project. Jesús Martinez Álvarez, the secretary general of the state of Oaxaca (SEGEGO, *secretaria general de gobierno del estado de oaxaca*), the state's second in command, had now been tapped by the governor to try to resolve the stalemate. When we met with Secretary Martinez in Oaxaca City in mid-January 2013, he appeared exhausted, and on his desk was evidence indicating why: a thick manila folder with "San Dionisio del Mar" written across it in permanent ink. Despite the file bulging with documents, the secretary shared that "even after two or three months of watching this conflict, we still do not have all of the information we need as a government . . . and we still have many competing interests."

The secretary explained that he believed that the amparo would be overruled in Mexico City. And he was convinced, at least rhetorically if not in his affective demeanor, that the conflict would be resolved soon. He thought this because, as he told us, the problem would be solved despite how many "mistakes" the company had made. Among these errors he included failing to install public works such as health centers and paved streets, which other companies had done elsewhere. Secretary Martinez felt confident that an agreement could be reached regardless of the involvement of "outside" organizations, which he called the "antieólicos." These antieólicos—a category he himself put in scare quotes—were those whom he believed were compelling people to action in Álvaro. He noted that his office had spoken with fishermen in the region and that they espoused faith in the government, though not, he emphasized, in the company.

Martinez explained in clear terms that he would not be willing to trample the rights of indigenous people in the isthmus. He was seeking a compromise if there was one to be found. "There has been error after error," he noted, including massive payments to municipal authorities who changed office every three years and, he went on, might not be trusted to properly utilize funds for the community's benefit. But the secretary saw a deeper omission as well, and that was a fundamental lack of awareness on the part of the company. "These mistakes have occurred," he said, "because they don't understand the mentality of the Huaves."

In an uncanny anthropological turn, Secretary Martinez posed the question in our direction. "Do you know anything about Huaves?" he queried. "Just type it into the internet. Originally, they're from Nicaragua, oppressed by the Zapotecs. They take their ceremonies very seriously." He referred to the "simple fishing life" of the Huave people, noting that it was "rudimentary" in form, utilizing "canoes and nets." (Whereas all the boats we had seen in use had motors and were usually made of metal, the secretary seemed to have a distinct image of ancient forms of subsistence, which he also seemed to believe could benefit from modernization and development.)[3] The secretary questioned whether fishing off the barra itself was even active; he was convinced that the antieólicos were simply using ikojts' fishing and ceremonial practices as a ruse to forward their own political agenda.

Secretary Martinez was confident that the issue would resolve itself within a few days and that he, a man who regularly worked sixteen-hour days, would not need to intervene further. By February, however, when we met with the secretary again, his "San Dionisio" folder had grown in size, and SEGEGO was fully enrolled in the negotiations among local protestors, company representatives, and municipal authorities. Political attunements were especially critical for state government officials like Secretary Martinez because each needed to find points of reconciliation; this entailed placing fault somewhere, but the location of that culpability was becoming increasingly nebulous.[4] The locus of blame was ever shifting: outside agitators, state officials' ineptitudes, malevolent refusers, ill-informed residents, and clueless corporate actors.[5]

More Police Trucks, Again

With the injunction in place and the barra still being defended in Álvaro Obregón, it was an inauspicious time to cross lines in the sand. Nonetheless, in the early morning hours of January 30, someone "claiming to be a biologist" attempted to enter the barra at the site of the barricade and was summarily and forcibly turned away. Locals at the scene reported that the individual had the insignia of Mareña Renovables on his materials. Following the confrontation that ensued, a representative from the secretary general's office appeared with six state police trucks carrying armed officers. A couple of short-lived arrests were made, a prelude to the interventions that the state would attempt next.

In a highly mediated press conference the following day, Mareña executive Jonathan Davis Arzac announced, "Investors have finally lost their patience" with the conditions confronting their development in the isthmus. The blame, Davis declared, should be attributed to the state government. "There is nothing that we have failed to complete [in terms of impact reports, legal obligations, or protocols]. . . . We want only to complete the project, and we call upon the authorities to apply the rule of law against those that have infringed upon it." He went on, "If the Governor of Oaxaca, Gabino Cué, does not guarantee stable conditions and the rule of law, we will have to withdraw our investment that would have gone to communities in the region." Davis asserted, "Only if the rule of law is applied will Mareña Renovables stay [in the isthmus] for we have already decided to go to another state in Mexico or perhaps another country altogether." And in a final flourish to underscore the consequences as he saw them, Davis reminded his audience that the forced departure of the Mareña project would set a "bad precedent" for the state of Oaxaca at both the national and international levels. Davis's threat to abandon Oaxaca and his call for the "rule of law" were clear requests for yet more police intervention. And Governor Cué, obviously concerned about divestment, seemed ready to oblige.

Just after darkness fell the following night the state police were again in Álvaro Obregón to break the barricade. After a brief skirmish on the sands near the entrance to the barra, the police withdrew to regroup. According to some present, they vowed that they would return with the army. Some reports claimed that two hundred police officers in riot gear were there that night, but photos from the confrontation suggest only a few dozen police were involved. News reports the next day stated that approximately forty protestors were there to defy state police, and all the protestors, some reports claimed, were "drunk." The resistance, unsurprisingly, refuted the claim of inebriated agitators. Whatever their numbers or condition, the state police did retreat from the scene, and protestors made off with several prizes of their own, including riot shields and clubs that had been acquired, they said, "when the police ran away."

The police were gone for the time being, but death threats against the resistance continued. Mariano, who had become an increasingly vocal and visible opponent of the wind power projects, told us that he received a death threat the night after the confrontation with state police. It came in the form

FIGURE 6.4. Riot shields, taken from the state police during the confrontation in Álvaro Obregón, overlaid with sticks used by protestors. Photo by José Arenas.

of a phone call, he said, which he believed came from a local gunman (*pisto-lero*), telling him to get himself out of Álvaro Obregón. According to others in the resistance, assassination plans were being devised for Rodrigo and Bettina as well.

Secretary Martinez continued voicing the government's position to the press, insisting that the real problem was a persistent "lack of information." He explained to us, "What you notice [in these conversations and meetings] is that these demonstrations are not against the project but rather that local fishermen and fishing cooperatives in the region do not have the necessary information." What was needed, as he saw it, was an initiative that would have a direct and immediate impact. The government needed to create jobs for the people of Álvaro and surrounding communities: short-term hard labor that was decently paid and beginning soon. The state government devised a temporary works program to clear overgrowth from irrigation canals on agricultural tracts around Álvaro and Zapata. Providing payment to the resistance in the form of a daily wage rather than in the form of bribes, would, Secretary Martinez believed, bring the confrontation to an end. The gesture was intended to demonstrate to local residents "that there is work that comes with the eólicos."[6] The secretary was sure that this would resolve the matter within the week. But it did not.

In the first week of February, Amnesty International Mexico issued an official statement denouncing human rights violations and threats against wind power activists across the isthmus. By the middle of February, *La Jornada* reported that sixteen international organizations had called upon the government and Mareña to cease violence against those opposing wind development in the region. As national and international human rights organizations were decrying threats, potential assassinations, and pressures against the antieólico collectives, Mareña fashioned a new financial proposal.

The company would, it explained, place all of the project's proceeds that were intended for community investment in a secure account that was accessible only to a *fideicomiso neutral* (neutral fiduciary power). The selected fiduciary executor would be someone of the highest moral standing; in other words, the executor would be immune to bribes. The fideicomiso neutral's task would be to guarantee that wind power proceeds that were intended for community development would be used only for *obras sociales* (social works/projects). The world-famous Oaxacan artist Francisco Toledo, known as el Maestro, was floated as a candidate to manage the funds and provide moral legitimacy. But the proposal foundered.

Municipal and state officials as well as candidates running in the upcoming elections continued to call on the government to intervene and protect investment in the region. In turn, the resistance in Álvaro announced that they were in the process of forming a community police force (*policia comunitaria*), believing that they had to protect themselves, autonomously, from the forces of the state and the company.[7] Miming the policia comunitaria form that had become increasingly present among residents in Mexican towns where drug cartels operated and threatened local life, the resistance in Álvaro—as the town's longstanding reputation would have indicated— was becoming less and less "governable" by the state apparatus. No better sign of this came than when voting booths were torched the following year.[8]

The Request and Retractions

The final days of 2012 rendered an insuperable challenge to the project: the elimination of its financial backing. The day after Christmas, the Interamerican Development Bank (IDB), the park's primary funding source, was presented with a notice requesting an ethical and legal audit. The official complaint, a "request," came from the Assembly of Indigenous Communities of the Istmo de Tehuantepec—225 members of local ikojts and binnizá indigenous communities. Because of its mandate to allow for independent review when social or environmental harms might be present in bank-funded projects, the IDB was obligated to look into the accusations that the assembly presented.

The complaint neatly recapitulated the critiques that the park had faced on the ground locally, in the courts, and in the bureaucratic mechanisms of the state for some time:

> The Requesters allege that the planning, construction and future operation of the Project has caused and may continue to cause social harms to their communities, traditional cultures and way of life.
>
> They allege that the construction and future operation of the Project will cause environmental harm to their land and livelihoods.
>
> The Requesters allege that they were not consulted and that the planning and other activities should have taken into account the communal land tenure, social structure and customs of the local indigenous communities.
>
> The Requesters also allege the physical safety of some community members has been threatened and harmed due to their opposition to the Project.[9]

Before a full-scale investigation could proceed, a preliminary inquiry was needed to test the general veracity of the complaint. The Indian Law Resource Center, based in Washington, DC, sent one of its attorneys to do so. Leo Crippa, who worked with the nonprofit organization, arrived in mid-February for a first survey of the complainants' claims.[10] We traveled with Crippa throughout the contested region, from Juchitán to Santa María Xadani to San Mateo del Mar, as he collected statements that he would compile to assess whether the request had merit.

On Crippa's list of complainants with whom he should meet was Filiberto, one of the more recent voices to join the chorus of opposition to the Mareña project. We met with Filiberto at his home in the little hamlet of Santa María Xadani. With Crippa taking assiduous notes, Filiberto confirmed that his branch of the resistance had only become involved after a group of fishermen were prevented from accessing the barra. Company agents, the fishermen had explained, demanded that they show their identification cards before they would be allowed to access the water and their fishing grounds. Both the fishermen and others in Xadani saw this as a dangerous precedent, portending a future when local residents would be forbidden from entering their traditional fishing areas because all access points would be privatized and occupied by the wind park. Crippa inquired about the impacts this might have on fisherfolk and others in the region.

"Returning migrant laborers," Filiberto explained, "depend on fishing to support themselves because there is no other income source." He worried aloud how people would maintain access to the waters that provided subsistence; he was equally concerned that the yearlong disruption of fishing during the construction phase would be deadly for the community.[11] Filiberto was unqualified in his assertions to Crippa throughout. "Fishermen," he said, "overwhelmingly oppose" the installation of the wind park. He estimated that given the size of the fishing cooperatives in the region, anywhere between six hundred and twelve hundred fishermen and their families would be affected by both the park's construction and the ongoing limits it might place on access.[12] "Fishing," Filiberto concluded, "is the informal job taken up by the most economically vulnerable." Crippa made his notes, and we were off again to hear similar testimony in San Mateo, a "traditional" ikojts village at the edge of the lagoon. Fishing, one man explained, "eso es el banco de la gente—la pesca" (that is the people's bank—fishing). The sea is the bank.

By the end of our time together, Crippa had become convinced that the request brought by ikojts and binnizá communities had merit and, he would argue, an investigation should be granted.[13] If the request were accepted, the IDB would be compelled to reevaluate the entire Mareña project. A full investigation would be carried out by the IDB's Independent Consultation and Investigative Mechanism (ICIM/MICI) in order to establish whether actions on the part of the bank were not in compliance with their stated policies, and if not, how and why.[14] In addition, surfacing any "direct, material adverse effects—potential or actual—that might impact the requesters," would be part of the investigative team's mission.[15]

Retraction of IDB funds would be devastating for the Mareña project: IDB loans totaled almost $64 million in financing. Also lost would be a good portion of the project's ethical credibility, which had been buoyed (for some) by the bank's endorsement. The request meant that Mareña's management was facing another magnitude of legal and fiduciary challenges. This was no longer the opinion of one federal judge in a sleepy town but rather the threat of a full-scale inquiry into the project's practices by those who controlled much of the park's financial flow.

In mid-July of the following summer, Crippa reported to us that the panel of investigators had declared that the request had been deemed eligible for

FIGURE 6.5. Fish drying, Juchitán de Zaragoza

an independent investigation. As Crippa sketched it, this would be a move "to determine the adverse impacts on the communities and noncompliance with the bank's policies." He would not be part of the independent investigation, he said, but he hoped that we would be able to speak to the investigators when the time came.

In the seven-month interim, the number of signatories to the request had almost quintupled, swelling to more than 1,100 names.

With their mandate in hand, the independent investigative committee was given a multistage set of tasks, carefully enumerated and with clear deadlines and deliverables. They would, in brief, (1) meet in person with requesters; (2) meet in person with representatives of the borrower and project developers; (3) visit the project site and areas of influence; (4) verify the observations, allegations, and facts underlying the request and cross-check them with other community members who were not party to the request; (5) meet with relevant federal government officials including the minister of energy; (6) meet with representatives of the Government of Oaxaca; (7) meet with the bank's Mexico representative; and (8) seek public or other official documents that might be relevant to the request. And finally, as part of their fact-finding exercise, the panel would obtain "reliable third-party information" pertaining to alleged harms. In the summer of 2014 we would provide some of that "reliable third-party information" in a detailed narration of what we had seen, heard, and learned in all our time in the isthmus as observant participants in the spiraling story of the Mareña project.

From the Financial District of New York City to
Juchitán de Zaragoza

Mareña's tale would not, however, be complete without the addition of one more voice. In most cases, this voice would have been powerful, even decisive. But in the winds of the isthmus, it was instead one last gasp. Andrew Chapman is a man who is very accustomed to executive boardrooms in Manhattan, but he had made the long trek to Juchitán to save his park. As the senior financial officer overseeing the Mareña project, Chapman had, at long last, taken it upon himself to try to find a solution. Perhaps more than anything, he sought to understand what had gone so wrong. Spending several weeks in a hotel room in the sweltering bustle of Juchitán, Chapman's presence in town was either the last stand or the last straw for the company. The community-policing initiative had just been declared in Álvaro Obregón, with explicit reference to the fact that its very raison d'être was the prevention of more wrongdoing by the wind project and its promoters. Chapman explained in one of his many press conferences that instead of community policing, what was needed "was negotiation and openness to dialogue." The community-policing initiative, as he saw it, worsened conditions for conversation. And this prejudicial stance and refusal to meet, talk, and negotiate, he explained, gave the impression that people were wholly aligned with the resistance when, as he saw it, the opposite was true. They were instead, he believed, "losing the opportunity to development and to economic autonomy."

Anyone observing the drama of the Mareña project, even from a casual distance, could not have missed Chapman's presence in Juchitán. He regularly hosted press conferences and spoke out whenever and wherever he thought he might be heard. We too knew that Chapman was in town, and so when we spotted a willowy white man standing on the side of the road that feeds in and out of Álvaro, we knew it was him. As we skidded to a halt, raising a plume of dust into the late-afternoon air, we hoped that he would talk.

Chapman was a frustrated man. He agreed to speak with us, his translator standing by his side, listening carefully. Chapman spoke only a few words of Spanish, which surely made his ability to communicate with local residents all the more limited. Over the blustering wind that marks the late afternoon, he shared that he had just been told that it was unsafe for him to enter the hamlet of Álvaro, even with police escorts. "I am upset," he told us, "that I can't even talk to these people." He lamented that force had come to replace dialogue. It was hard not to sympathize with Chapman. He spoke openly, shouting through gusts of wind. He was a man longing to be heard. "My job is to go in there and try to open a dialogue and to go listen. But I can't do that with threats of violence. If it's safe to send my people in, I'll send them in. . . . But, you know, the only way to change minds is to listen to people. But if you're not allowed to listen to people, what do you do?" He threw his hands up in hopeless defeat. "We've got this project that I really believe is good for the planet, good for the region, good for the people down here." His reasoning was compelling.

> I mean, you can't help but be stunned by the beauty of this place. And then you see how the people are *living*. And I'm trying not to just impose my American values here, but I don't think lousy medical care is a good thing, that lousy schools are a good thing. . . . So if you can funnel resources into these communities to improve those services, imagine where they could be in five or ten years. They can still be fishing the lagoons, but they'd have basic stuff, like electricity that is continuous, like transportation, like schools. . . . It may sound very idealistic, but that's actually what we're trying to do. And to be confronted with this violence and with people who are essentially lying about what we're trying to accomplish . . .

Chapman then went quiet, his eyes focusing into the distance. He was not just exasperated, it seemed, but utterly spent. As to whether he had any patience left, he replied, "not much." Finding a final lungfull of air, he finished his thoughts. "I just find it frustrating and sad, and the consequence is that the investor group that I represent . . . they're sitting in their offices, and they can put their money here, they can put their money there. And they're just going to say to themselves, 'Why? I don't need these problems. I'm not actually in the business of saving the world, I'm in the business of earning money for my fiduciaries. And I need to do that in a low-risk way.'"

But this park had become, in no way, "low risk."

Chapman departed Juchitán a few days after we spoke to him on the road outside the town that he could not enter. He had already publicly admitted that he "did not want to go on in these conditions," and other Mareña senior officials—among them Sergio Garza and Jonathan Davis Arzac—had told Secretary General Martinez, "We are leaving because of all the trouble."

In mid-February, the secretary general convened a final *mesa de dialogo* (discussion session) with Mareña representatives as well as three hundred members of the community in Álvaro and its surrounding hamlets. Complaints about state police intervention, accusations that the company had not adequately informed fishermen of their rights of passage, and the impacts

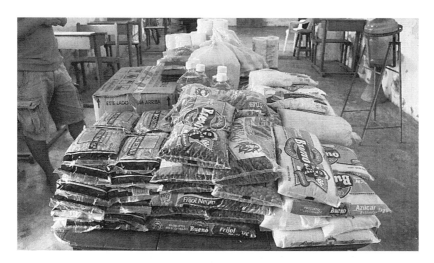

FIGURE 6.6. Supplies gathered for the humanitarian caravan, Juchitán de Zaragoza

of the park were again key issues expressed by the residents in attendance. More critiques came regarding political parties in the region and COCEI's operations in particular.[16] The government at all levels, state and federal, they felt, had neglected the binnizá population. The secretary general, was, as one report put it, "pinned to the wall," and he agreed that the state police would cease to intercede. He had become convinced that police raids on the town would never solve what he now plainly saw as "social issues."

The secretary general then packed his bags, promised to respect the wishes of the community, and he was off.[17] The little hamlets around the sandbar were now, depending on one's perspective, either abandoned once again or finally left in peace.

A few days after, the resistance began gathering supplies for "a humanitarian caravan" that would make its way from Juchitán to Álvaro Obregón. Convening in Juchitán, several hundred people made their way to Álvaro to deliver the mission. An aging pickup truck was filled with provisions for those still occupying the barricade at the barra: packets of pasta, jugs of oil, bags of sugar, coffee, matches, masa, and several packages of cigarettes. A thousand people showed up at the hacienda headquarters in Álvaro Obregón to greet the *caravana*.[18]

The day after the caravana humanitaria arrived in Álvaro Obregón, the Mexico City newspaper *La Jornada* ran the story that Mareña would not be built. Many thought it was true. Others saw it as a ploy and waited for the next threat to come.

Speculators

Mareña's disappearance brought a raft of speculation about what would come next. Many in the resistance in Álvaro, Juchitán, and San Dionisio thought that they had dealt a decisive blow to the company. Others were convinced that the project would be back, as had happened so many times before with other projects, "posing under another name." Rosa Rojas, a reporter for *La Jornada*, explained to us her theory that Mareña would sell its rights to the substation—the 396 megawatts it had bid on and for which it now had proprietorship through CFE. Other observers, from those in the resistance to state officials, were convinced that the project would be relocated inland, away from the barra and the trouble at the edge of the sea. Juchitecos averred that the project would probably move toward them, to a site that had been contracted earlier by Preneal. Sergio, for one, theorized that Mareña would move to El Espinal, where Preneal had also contracted land, and that it was already in the process of changing its name to Energía Eólica del Sur. Speculations about the wind park's fate were astute. After six years of negotiations, the wind park that used to be Mareña Renovables—now called Energía Eólica del Sur—began construction farther inland near the communities of Juchitán and El Espinal. The groundbreaking ceremony held in November 2017 featured the governor of Oaxaca as well as officials from the Mitsubishi Corporation; a handful of months later, the park faced an amparo issued by Mexico's Supreme Court. In an uncanny return, the project was found to have violated the rights of affected indigenous residents for failing to provide "free, prior and informed consent." In an equally uncanny turn, that amparo was then summarily rescinded and work on the park began anew.

Twice killed, the wind park that never was may yet be resurrected. But that remains to be seen.

Ending, without Wind Power

What wisdom is to be found in the tribulations of a wind park that was meant to provide a measure of respite from the Anthropocene but that instead seemed to threaten indigenous sovereignty and endanger the ecosystems that renewable energy is supposed to protect? It would be easy to say that the Mareña project collapsed because its directors had not adequately accounted for the history of the region where they aimed to build. They had failed to

FIGURE 6.7. Poster appearing in the isthmus after the announcement that the Mareña Renovables project had been terminated. Photo from the Asamblea de los Pueblos Indígenas del Istmo de Tehuantepec en Defensa de la Tierra y el Territorio, January 2014.

see, or willfully ignored, the proud local chronicles of insurrection. They had also failed to see, or willfully ignored, the record of abandonment and neglect that the area had suffered at the hands of the government and how corporate actors and transnational finance had sown deep misgivings nearly everywhere in this part of the world.

It would also be easy to say that the Mareña project collapsed because its management had not adequately accounted for the future of the region where they aimed to build. They had failed to see, or willfully ignored, that the priorities of residents might be to see their grandchildren gathering wealth from the sea. Or that monetizing land in the form of rents and bribes would never provide the security or sustenance that came in the body of a fish. Industrialized land and blocked passages to the sea would seem to portend a future full of money for some, but not for others. Foreign products, whether turbines or imported goods acquired only with cash, provided an unpromising future at best.

If the expectations of corporate actors did not coincide with those of local communities, there were many factors that produced that atmosphere. There were, without doubt, local caciques who manipulated company representatives and lined their own pockets, just as there were corporate representatives who, until it was too late, chose to believe they were in the right or refused to see any possibility otherwise. Self-interest drove deep into the debate, certainly not everywhere or for everyone, but for too many. Communicative channels did break down, iteratively, over time, and then festered until those conversations became unremediable. Hopeful potential became buried in unfulfilled promises. But if these breaches were especially great, they are nonetheless indicative of hurried development that is guided by privatization and profit and indulged by governmental authorities seeking to oblige companies to improve the precarious infrastructures of schools and health centers, paving and street lights, in the places where the state cannot or chooses not to do so.

The bribes alleged to have been distributed by the company, the faulty contracts, the gifts of trucks, the jailed guardians of the barra, the assaulted inconformes, the suspect political machinations of local, state, and federal leaders, all might have been overlooked. They might even have been taken as the norm: a recapitulation of the old, familiar habits of capitalism, caciquismo, and corruption. The vigilantly crafted environmental impact reports; the relocation of the park's footprint; the monetary benefits distributed to local communities; the plans for soccer fields, health centers, basketball courts, and community centers; the capital investment; the infrastructural projects; the road paving; the handful of jobs; the construction contracts and the union work, all might also have been taken as fair compensation for a project that would, ultimately, provide clean power.

It might have worked as it was intended were it not for an unforeseen and vital relationship between fish and people.

The big wind park was born from the logic that its global climatological good would ultimately correspond with the ecological, economic, and social worlds that comprise human and other-than-human life across the isthmus. Its power was meant to be an antidote to the Anthropocene. But failures of attunement prevailed. The story of Mareña, unfolding in times of ecological precarity, demonstrates the ways that environment and energy are carved in parallel form. Ecologies of energy are then mutual acts created between forces and materials, as well as humans and their others, each of which are bound up in the wind made electric.

Joint Conclusion to *Wind and Power in the Anthropocene*

CYMENE HOWE AND DOMINIC BOYER

Aeolian Politics, Aeolian Futures

We went to the Isthmus of Tehuantepec as ardent supporters of renewable energy transition, and we left with that conviction intact. Wind power (alongside solar power, tidal power, geothermal power, and biofuels) has an important role to play in reducing greenhouse gas emissions and decarbonizing electricity generation. But we also returned to the United States with a more skeptical view of renewable energy's capacity to guarantee salvation from climate change let alone the Anthropocene. Renewable energy has a necessary but insufficient role to play in a process that will amount to a refashioning of the civilization(s) that brought us to our present ecological and political conditions. What our field research on Mexico's aeolian politics and the ecosystemic limits of wind power taught us above all is that it is all too easy for renewable energy development to occur with little or no social, political, or economic transition attached to it. It is both possible and common to build wind parks firmly within a model of resource extraction that is typical of global fossil fuel and mining industries. We have offered extensive documentation of such wind development in our Mareña (*Ecologics*) and La Ventosa (*Energopolitics*) case studies—where attempts to capture the wind resulted in failures, both human and other than human. We have also shown in the case of Ixtepec (*Energopolitics*) that other development models exist, even if they are being actively resisted in Mexico. Where human desires for energy are not in balance with their ecosystemic context, as we see across the

Ecologics volume, there is little hope of remediating climates either locally or globally.

Taken together, the three case studies we have followed in *Ecologics* and *Energopolitics* demonstrate the turbulence surrounding renewable energy as the world awakens to the Anthropocene. They tell stories that are specific to Mexico and yet also exceed national boundaries. Carbon politics, finance capital, global industry, consumerism, and a lack of environmental protections have laid deep infrastructural grooves and have largely drawn aeolian politics into their orbits. Thus, the win-win-win visions of green financiers, entrepreneurs, and developers who promise that climate change can be reversed while maintaining everything else about the modern world, especially economic growth and a positive return on investment to shareholders, show a stubborn reluctance to abandon the structural deficits of carbon-based modernity. Those imaginaries are shared to a great extent by Mexican and Oaxacan politicians and technocrats who, steeped in neoliberal certainties and petropolitical anxieties, yearn for foreign direct investment to extend and improve the biopolitical functions of governance in the form of health, security, and prosperity. Some even believe that wind power can help to fulfill delayed or abandoned plans to bring, at long last, the isthmus into the nation, not as a repartimiento vassal but as a vigorous organ of the *mestizaje* national body. Local leaders and asambleas, elected and unelected, are likewise drawn toward the biggest influx of international attention and activity the isthmus has experienced since the mid-nineteenth century. Some fight for local or indigenous autonomy and sovereignty against the encroachment of *megaproyectos*, others pursue windblown wealth to further dreams of better jobs for their children or the accumulation of capital and leverage or for the opportunity to extend and deepen their networks of influence. It is not only in Mexico that dreams of aeolian futures are paradoxical; what are heavenly images for some are nightmares for others.

This is only to speak of the anthropolitical dimension of aeolian politics. We must also consider the Anthropocene trajectories of birds and bats and fish, the machinic life of turbines, the grid, and trucks, the unruly howl of el norte, and the gentle breezes of binisá. Aeolian politics is always already more than human even if the ecological interdependency of human and nonhuman potentials is largely ignored in standard treatments of wind power. It is for this reason that we have created a duograph to offer not only an ethnographic division of labor in its coverage of the three studies but also an analytic division of labor that allows us to pursue, with better depth and peripheral vision, both the mapping of anthropolitical enablement and the mesh of human-nonhuman relationality that is often allowed to drift into the background of reckoning with the Anthropocene. Questions of wind and power circle each other in the Isthmus of Tehuantepec—How can the fierce northern winds be harnessed? With what machines? To what end? Benefitting whom? Displacing whom? Earning what? Killing what? For how long? And with what consequences? We have likewise sought to let the analytics of wind and power speak to each other in this duograph, probing their potential to remake and unmake the Anthropocene. Enablement is always relational: some complex of forces, things, and events begetting others. Relations, for the same reason, always enable. The riddle of the Anthropocene is what mesh of relations and actions will allow us to disable the reproduction of the present while being present in the production of a future. For those who wish to solve that riddle, we must attend to both human politics and all the other relations and forces that make those politics possible.

An earlier version of our duograph was titled *Winds of Desire* because everywhere we turned in Mexico, we found people wishing for the wind to deliver something: money, electricity, influence, legitimacy, prosperity, development, power. At times, desire cloaked itself in mathematics, rationality, and common sense. At other times, it reveled in naked hallucination. Those who desired were rarely satisfied with what the wind had already delivered to them. What desire always accomplishes best is the propagation of more desire. Here, at the end of a project that has been nearly a decade in the making, we are asking ourselves what it is that we wish from wind power. It turns out that our object of desire is also elusive and receding. Still, we are drawn toward it: we want better aeolian politics oriented toward achieving better aeolian futures.

Our final report to the National Science Foundation listed the following findings and recommendations based on our research:

> The field research for NSF #1127246 yielded several important findings and recommendations that will contribute to more positive development outcomes in Mexican energy transition in the future. (1) The dominant development model prioritizes the interests of international investors and developers and local Isthmus political elites over other stakeholder groups, especially the regional government and non-elite Isthmus residents. (2) The dominant development model has reinforced hierarchy and inequality in Isthmus communities through unequal distribution of new resources like land-rents. (3) The development model has generated significant polarization in

Isthmus communities regarding wind parks and undermined trust in government and industry. (4) The financial benefits from land rents are currently primarily being directed toward luxury consumption by elites. (5) A majority of Isthmus residents appear to favor wind power development were its financial benefits to be more equally distributed. (6) Project findings suggest that the Mexican government needs to reevaluate its development model to guarantee (a) that entire communities and not simply elites are involved in project design and implementation, (b) that mechanisms be developed to guarantee that wind power development yields consistent and significant public benefits, and (c) that regional governments receive sufficient federal funds to develop a regulatory agency with the authority to guarantee that wind power development is truly transparent and beneficial to all stakeholder groups.

To put this in less muted terms, in our view, there will be no "renewable energy transition" worth having without a more holistic reimagination of relations in which we avoid simply greening the predatory and accumulative enterprises of modern statecraft and capitalism. In this respect, the record of Mexican wind development thus far does not inspire much confidence. The model of wind development that currently dominates the isthmus has been very effective at building wind parks, but it has done almost nothing to disrupt the toxic kinds of relatedness that made it necessary to build wind parks in the first place. It has left wind power in the thrall of finance capital, state biopolitics, and energopolitics; parastatal utilities and infrastructure; PRIismo, caciquismo, consumerism, and many other -isms besides. The case of Mareña Renovables (in *Ecologics*) came to absorb and reflect all these conditions and in so doing was stalled out of existence. In failing to account for local histories and imagined futures, and in repudiating local worries about environmental harm, Mareña's potential to provide climatological remediation and reduce greenhouse gas emissions was drowned among the fish. With the Yansa Ixtepec project (in *Energopolitics*), we do find a scrappy DIY prototype for a better aeolian future, one that seeks to harness wind-generated electricity to help a rural farming collective to better guarantee their own autonomy and futurity while still contributing to the global cause of decarbonization. Yansa Ixtepec has flaws to be sure—its benefits will not extend far beyond the collective, and it requires a grid and a failing parastatal electrical utility to pay its rents—but if the project is ultimately thwarted, Mexico will miss its best chance to connect the heady ambition to be a global leader in clean energy development with the interests, hopes, and worldviews of people living in places where the wind is strongest. In the end, we need not just new energy sources to unmake the Anthropocene, we need to put those new energy sources in the service of creating politics and ecologics that do not repeat the expenditures, inequalities, and exclusions of the past.

We will conclude with an appeal for more collaborative anthropology in every sense of the term. We need more anthropologists working together and working with other humans and nonhumans on the problems that matter most in this world. Those problems, like energy transition, are complex, massively scaled, and very often ill suited to critical and activist engagement by individual researchers. As scholars, we will better understand our present dilemmas and possible paths forward if we work together, whenever possible drawing on varying but complementary skills and forms of expertise in the pursuit of responses. As beings living on a damaged planet, what we already understand is that none of us can exit the Anthropocene on our own. The hyperindividualism of the past three decades, the capitalist empire building of the past two hundred years, the Northern privilege of the past five centuries, the monotheistic patriarchy of the past two thousand years, the agrilogistics of the past ten millennia—all of this, everything, will have to be remade if a global humanity is going to be reborn that will not be actively, constantly destroying its lifeworld and the lifeworld of the majority of the earth's species. This project will be utopian in the sense that it will have to make a world that has not yet existed. It will be revolutionary in the sense that it will not be accomplished by technology, or markets, or violence, or anthropocentrism, or any of the other behaviors and attitudes that brought us here in the first place. It will be a project accomplished by humans who can accept their own diminishment of importance and entitlement relative to their nonhuman neighbors and by those who are willing to work collaboratively to restabilize the vital systems of geos and bios on this planet. These are the politics, aeolian and otherwise, to which we should commit ourselves, these are the futures worth having.

Notes

JOINT PREFACE

1. See Lynch 1982; Price 2016.

2. For more information on these partnerships, see Ethnographic Terminalia, http://ethnographicterminalia.org; "Anthropology of the World Trade Organization," Institut interdisciplinaire d'anthropologie du contemporain, February 12, 2008, http://www.iiac.cnrs.fr/article1249.html.

3. But here, as in other respects, we find the aforementioned collaborative partnerships trailblazing. See, for example, Matsutake Worlds Research Group 2009; the exhibition catalogs and zines produced by Ethnographic Terminalia, http://ethnographicterminalia.org/about/publications; Abélès 2011.

4. See, for example, Boyer and Marcus, forthcoming.

INTRODUCTION

1. For anthropological and other social scientific approaches to the study of energy, see, for example, Boyer 2014; Daggett 2019; Howe 2014, 2015a, 2015b; Howe and Boyer 2016; Howe, Boyer, and Barrera 2015; Hughes 2017; Krauss 2010; Love and Garwood 2011; Mason and Stoilkova 2012; Nadaï 2007; Nader and Beckerman 1978; Nader 2004, 2010; Pasqualetti 2011a, 2011b; Pinkus 2016; Scheer 2004; Strauss, Love, and Rupp 2013; Watts 2019; White 1943; Wilhite 2005; Winther 2008; Winthereik 2018; Wolsink 2007.

2. Throughout this text I use the terms "we" and "our" with different intentions that I believe the reader will find clear in context. In some instances, "we" (or "our" or "us") is in reference to the collaborative research team of two. At other times, the "we" refers to those of you who are reading this text and therefore engaging in a conversation about the issues that are included here. And finally, there are instances where "we" is meant to speak of and to a grander category of human beings. The latter usage of "we/our" is clearly universalizing in some ways, indexing "all of humanity."

However, my intention is not to presume that all humans are equally positioned to act or respond to the environmental dilemmas that are the context for this project nor to suggest that all people—past, present, or future—are their root causes. Instead, I want to draw attention to humans as a species that has, through some of its agents over time, manipulated earth systems and "resources" to the point where it is now unclear whether a collective human effort will be able to control the environmental consequences that have come from carbon incineration and other earth-altering practices. Above all I want to emphasize that "we" is always a heterogeneous human.

3. For anthropological work on global warming, climate models, climate politics, and climate impacts, see Barnes et. al. 2013; Crate and Nuttall 2009; Edwards 2013; Henning 2005; Hulme 2011; Klein 2015; Lahsen 2005; McNeish and Logan 2012; Monbiot 2009; Oreskes and Conway 2011; Rhoades, Zapata, and Aragundy 2008; Roncoli, Crane, and Orlove 2009; Strauss and Orlove 2003.

4. By "subjunctive future" I resort to a (rarely used) grammatical form, the future subjunctive (available in Spanish and other language systems) to indicate what might be or that which could be were a certain set of predecessor events and qualities to unfold prior to that future moment being indicated: a hypothetical future action. In contemporary usage, the future subjunctive has been subsumed into the present subjunctive and appears only rarely (for instance, in literary or legal documents). However, here I want to underline both the future (temporality) and the subjunctive (possibility). I contrast this with Kim Fortun's "future anterior" (2001, 353). For Fortun, the future anterior is a formula for prefiguring the future by assessing the past (and thus aspiring to a better and better-understood future), whereas the future subjunctive is less sensitive to the past than it is to the present-cum-future.

5. I use the term "fisherfolk" to designate both those who actively fish and those who process and vend the fished products. We never encountered a fisherwoman in the isthmus during our research, though women were very involved in fishing as a livelihood. The *séptima* neighborhood—a working-class barrio where many Juchitecan fisherfolk live—is buzzing with women cleaning, drying, and selling fish in the predawn morning.

6. Social scientific work on infrastructures has been burgeoning. See, for example, Anand 2017; Appel 2012; Barnes 2014; Bowker et al. 2010; Carse 2014; Gupta 2015; Harvey and Knox 2015; Howe, Lockrem et al. 2015; Larkin 2013.

7. Beyond privately owned parcels of land, two forms of land tenure serve as important social forms in the isthmus and in Oaxaca more generally—*bienes ejidales* (or ejidos) and *bienes comunales* (or *comunas, comunidades*). Ejidos, a product of the Mexican Revolution, allow mestizo peasant farmers to collectively maintain and manage a communal estate, usually for the purposes of farming; members are referred to as *ejidatarios*. In the 1990s ejido collectives were able (and sometimes encouraged) to privatize land parcels, converting them into private properties with deeded owners. Bienes comunales are likewise collectively managed communal estates, but they are recognized as having belonged historically to indigenous peoples, gathered together as an *asamblea* or *comuna*; members are referred to as *comuneros*. Bienes comunales maintain a governing structure that calls upon the community's overall

membership (the asamblea) to vote in matters of land disposition. The commissariat (*comisariado*)—composed of a president, secretary, and treasurer—is charged with the administration of proper procedures and decision making, and they are in turn supervised by a *consejo de vigilancia* (oversight committee), comprising three comuna members, with elections taking place every three years. For more on ejidos and bienes comunales, see Cornelius and Myhre 1998; Castellanos 2010.

8. Remoteness is, of course, relative. For those communities being impacted by the extraction of fossil fuels and those laborers who work in the industry, extraction can be intimately felt on a daily basis.

9. This research was a collaborative project with Dominic Boyer—beginning in 2009 and concluding in 2013—that investigated the political and ecological dynamics of wind power development in Oaxaca, Mexico. For more on collaborative analytics in anthropology see Marcus 2018 and on authoring and writing in anthropology see Wulff 2017.

10. In 2013 the Mexican state undertook energy reform measures, revising its seven-decade-long commitment to nationalized oil production and ending Pemex's role as the sole owner and operator of the country's fossil fuel assets. In spring 2017 an Italian company was the first international operator to drill in Mexican waters, and it is expected that oil production will increase in the coming years. On Mexican energy reform, see the International Energy Agency report, "Mexico Energy Outlook."

11. See Booth 2010, for example.

12. On wind resources in the isthmus, see Almeyra and Alfonso Romero 2004; Alonso Serna 2014; Aiello et al. 1983; Borja Díaz, Jaramillo Salgado, and Mimiaga Sosa 2005; Caldera Muñoz and Saldaña Flores 1986; Elliott et al. 2003; Hoffman 2012; Sánchez Casanova 2012.

13. In June 2016 the US, Canada, and Mexico agreed that they would jointly commit to 50 percent noncarbon fuel sources (for electricity generation) by 2025; this represents a significant upscaling of Mexico's original formulation. Note that "clean" energy sources in this context include not just renewables but also nuclear energy, carbon capture and storage plants, and energy efficiency. Under that definition, 37 percent of North America's electricity in 2015 came from clean energy sources (Eilperin and Dennis 2016). Just 22 percent of Mexico's electricity generation in 2014 came from nonfossil fuels, according to its government, though the country has pledged to raise that to 34 percent by 2024.

14. La Asociación Mexicana de Energía Eólica, A. C.

15. The number of Mexican households that could be served by this quantity of wind-powered electricity is difficult to predict. Calculations of household electricity are complex and contingent on several factors. Electricity demands differ from state to state according to climate, habits, and installed devices. For one study of both urban and nonurban households in Mexico derived from the Encuesta Nacional de Ingresos y Gastos de los Hogares 2008 (ENIGH), see Cruz Islas 2013, 198.

16. Or, for that matter, any other environmentally disruptive extractive practices exercised in the name of modernity and growth. See Bebbington 2009; Galeano 1997; Gudynas 2009; Johnson, Dawson, and Madsen 2007; Liffman 2017; Turner 1995;

among others. On waste see, for example, Alexander and Sanchez 2018; Alexander and Reno 2012; Gabrys 2013. On the petropolitics of oil and its afterlife specifically, see Behrends, Reyna, and Schlee 2011; Breglia 2013 (Mexico); Coroníl 1997 (Venezuela); Klieman 2008 (historic, Congo); Sawyer 2004 (Ecuador); Mitchell 2011.

17. This can also be taken as a sign of cynical reason, or what Peter Sloterdijk (2014) calls "enlightened false consciousness": people are equipped with knowledge but refuse to act accordingly.

18. For examples of oil and crises, see Bini and Garavini 2016; Dietrich 2008; Mitchell 2011; Love 2008. However, from my point of view, questions of energy transition in the Anthropocene provide a deeper impetus to enact and live energy/environment "otherwise." The environmental precarity of the present—in its global sweep and interlinked ecocrises of melt, seawater rise, and climatological decay/precariousness that are scientifically proven—suggests a unique condition for energy as well as encounters with and articulations of environment.

19. Changing our collective forms of energy is, from my point of view, an unqualified necessity, and this book is certainly not an argument against renewable energy nor against wind power as an important node of that apparatus. The question, rather, is how transitions can be undertaken with more care and attention to potential harm than has often been the case in the past.

20. Kathryn Yusoff describes this potential as the "extinguishment of the late Holocene human subject" (2016, 5).

21. Again, I want to bracket the grand human "we" here in the recognition that not all humans have contributed equally to, nor will suffer equally with, anthropogenically induced changes to the earth system (see Davis 2010 for an excellent, related discussion). There has been a tendency, in discussions about the Anthropocene, to imagine "future humanity" as a way to erase contemporary social differences and inequalities, including climate racism, as Kathryn Yusoff has pointed out (2016, 2). I do not want to rehearse that elision here, but I do want to focus on modulating the false separation of human and nonhuman survival and extinctions.

22. See, for example, Scranton 2015.

23. Humans as a "weedy species" (Wake and Vredenburg 2008) seems to be a more and more resonant designation, especially in the context of "ruins" and "blasted landscapes."

24. Wind machines (to test aerodynamics, for example) or fans (for cooling) are instances of human-generated wind, but their fundamental property continues to be (gaseous) movement and interaction. Unlike solid (minerals, coal), liquid (water), or viscous (oil) resources, wind is only generative when it is in motion. It is contrastatic.

25. The ecology of relationships builds from Descola 2013a, 5. In seeking to avoid a strict division between ontological and phenomenological being, I am thinking of productive pairings of the two. See, for example, Bennett 2010; Braun and Whatmore 2010; Chen 2012; Descola 2013a, 2013b; Jasanoff 2010; Massumi 2009.

26. For a range of more recent interpretations as to what constitutes "Nature's" end or its radical reformulation, see, for example, McKibben 1989; Latour 2004a.

27. Claude Lévi-Strauss and allied structuralists come to mind, but the human sciences have been in a more protracted discussion over the definitional qualities of nature/culture for far longer.

28. See, for example, Strathern 1980, 1992.

29. See for example Dipesh Chakrabarty's influential 2009 essay, "The Climate of History: Four Theses." His theses are (1) "anthropogenic explanations of climate change spell the collapse of the age-old humanist distinction between natural history and human history"; (2) "the idea of the Anthropocene, the new geological epoch when humans exist as a geological force, severely qualifies humanist histories of modernity/globalization"; (3) "the geological hypothesis regarding the Anthropocene requires us to put global histories of capital in conversation with the species history of humans"; (4) "the cross-hatching of species history and the history of capital is a process of probing the limits of historical understanding." For further perspectives on the Anthropocene, see, for example Steffen, Crutzen, and McNeill 2007; Steffen et al. 2015.

30. In his 2014 distinguished lecture delivered at the American Anthropological Association meeting, Bruno Latour saw the advent of the Anthropocene, and scholarly work on it (1) to focus upon "human agency" as its central tenet, (2) to explicitly conjoin the "physical" and "social" sciences, and (3) to raise moral questions of responsibility (or as Haraway would have it, response-ability), all of which anthropology has been doing all along (Latour 2014, 2–4).

31. In fact, it would be impossible to narrate a history of anthropology without accounting for the significant role of nonhuman animals in ethnographic work from the inception of the discipline to the present. Early examples include Lewis Henry Morgan (1868) on the American beaver (a more naturalist account) or his account of Iroquois phratries (wolf, bear, and turtle, for example) and Boas's research on seal-hunting practices among Inuit peoples on Baffin Island (1883). While some human/nonhuman animal encounters are described in more programmatic terms (such as hunting), anthropology has represented a wide range of animal-human lifeways. Think of Cushing and Benedict on Zuni animal tricksters, Mauss's (1979) explicit ecological frame for his "social morphology" hypothesis, or Rappaport's (1968) deeper ecological approach concerning humans and their eco/animal. Douglas's 1957 discussion of human/animal relations among Lele peoples, for one, presages many contemporary discussions of human/nonhuman relationality. She writes that for Lele, one of the defining principles of animals is "their own acceptance of their own sphere in the natural order. . . . Most run away from the hunter, . . . but sometimes there are individual animals which, contrary to the habit of their kind, disregard the boundary between humans and themselves. Such a deviation from characteristically animal behavior shows them to be not entirely animal, but partly human" (1957, 48–49).

32. Social scientists concerned with other-than-human life as well as those committed to more deeply investigating the ways that inanimate materials shape human (or nonhuman) beings are many and growing. See, for example, Alaimo and Hekman 2008; Candea 2013; Coole and Frost 2010; de la Cadena 2015; Franklin 2007; Hartigan

2015, 2017; Hird 2009; Kirksey 2014; Kohn 2013; Lowe 2010; Myers 2016; Nadasdy 2007; Nading 2012; Paxson 2008; Porter 2013; Raffles 2010; Stengers 2010; Stewart 2011; Tsing 2012, 2015. In the humanities, see Wolfe 2009 among others.

33. For biology, see, for example, the paradigm-altering biological research of Lynn Margulis (1970); John Hartigan's excellent work on genomics, science, and racism in Mexico (2013). Regarding physics, Karen Barad, a theoretical physicist and feminist philosopher, develops the concept of "agential realism," which serves as an epistemological and ontological framework to center on the nature(s) of materiality and those relationships to discursive forms. The intention is to reform both "agency" and "realism," to underscore how human and nonhuman factors intervene in how knowledge is produced. In other terms, agential realism tries to move beyond the usual dyadic interpretation that distinguishes between social constructivism and conventional forms of realism (2003). Thus, agency, for Barad, "is a matter of intra-acting; an enactment, not something that someone or some-thing has."

34. See Alaimo 2010, 2016. Also see Haraway 1996.

35. The literature on actor-network theory is too massive to fully include here. However, for a comprehensive, chronological list of ANT texts and responses, see "ANT Resource," Centre for Science Studies, Department of Sociology, Lancaster University, http://www.lancaster.ac.uk/fass/centres/css/ant/ant.htm, last updated 2000.

36. Kim Fortun warns, for example, of what she calls the "Latour effect" in anthropology and science studies: that is, a singular focus on practices of expertise and actor networks in late industrialism that does not account for the material and social matrix of the toxic and inhospitable environments that make up people's lives today (2014).

37. See Barad 2003, 806–7.

38. On "worlds" and "worlding," see, for example, de la Cadena 2015; Viveiros de Castro 1998.

39. In "Posthumanist Performativity," Barad (2003) is responding to theorists of performativity, in this case Judith Butler, but by extension a whole oeuvre of post-structuralist work on discourse and the hailing of iterative linguistic performance that has derived (largely) from the work of linguist J. L. Austin.

40. Many alternative designations for our current age have been proposed in recent years: "Eurocene" (Grove 2016); "#Misanthropocene" (Clover and Spahr 2014); "Naufragocene" (Mentz 2015); and perhaps best known (currently), Donna Haraway's "Chthulucene," a period of "collaborative work and play with other terrans," where "flourishing occurs across assemblages of intra-active multispecies life, that includes more-than-human, other-than-human, inhuman, and human-as-humus" (2015).

41. See Strong et al. n.d. for citation practices and female authorship in cultural anthropology.

42. Compare to Tim Morton's "agrilogistics" (2016), which locates roots of the Anthropocene in the advent of agriculture and its material and ideological force beginning about ten thousand years ago. The Plantationocene indexes a more recent period of colonial expansion and its continuing effects.

43. In 2016, after seven years of study, an eminent group of scientists and scholars called The Anthropocene Working Group—composed of geologists, engineers, paleobiologists, geographers, historians, and philosophers among others—declared that the

world had entered a new geological epoch called the "Age of Man." The panel reported that biospheres, lithospheres, hydrospheres, cryospheres, and atmospheres everywhere on earth contained the imprint of human activity, including radioactive debris, plastic tides, displaced soil, and increased methane and carbon dioxide.

44. For a useful overview of Anthropocene "sources," see, for example, Bonneuil and Fressoz 2016.

45. Allochronic time occurs in a different geologic time. Anthropology itself has struggled with such allochronicities, namely the mistranslation of space into time. As Johannes Fabian (1983) has famously pointed out, the discipline has crafted reports that deny the coevalness between the ethnographic subject and her ethnographer. "Savages" could be temporally displaced, cast back in time as primitives, their worlds made static, largely because of their remoteness from "civilization." Fabian's formulation of allochronic, asynchronous time in the context of the Anthropocene may be worth revisiting as a way of recalibrating human time into geologic sync with nonhuman materials and beings.

46. See, for example, Kolbert's *The Sixth Great Extinction* (2014).

47. An emphasis upon periodizations of the Anthropocene also speaks to Chakrabarty's (2009) theses where historical time frames, or periodizations, that separate human from natural history come under critique. Or we can think about Tim Morton's admonition that while the Anthropocene time line may be "fuzzy" (Was it the advent of agriculture? Was it the industrial revolution? Was it the Great Acceleration?), we can nevertheless find an operative set of coordinates, for it is clear that it did not start 1.3 million years ago (2013; 2016).

48. Yusoff (2013a, 781) writes that in the Anthropocene, with humans as geomorphic agents, "new understandings of time, matter, and agency" accrue for the human as "a collective being." Through the immersion of humanity in geologic time, she suggests a move away from (simply) biological life courses to instead "a remineralisation of the origins of the human" as well as a shift in human time scales to stretch toward the horizons of the epochal and species lifescapes.

49. See, for example, LeMenager 2014; Zalasiewicz 2012.

50. Povinelli 2016, 8–9. In *Geontologies*, Beth Povinelli makes the argument that "geontologies" have long been here with (and of) "us" but that the conditions of the Anthropocene may be surfacing that fact to some human beings (often settler-colonialist societies), whereas many indigenous peoples, like those who have become Karrabing, have in fact recognized this ontological reality all along (see especially chapter 2). The separation of life and nonlife, she goes on to state, is also a technique of settler colonialism that has historically been used to debase indigenous ontologies and cosmologies that take nonlife beings as sentient. See also de la Cadena 2015.

51. The term "Plantationocene" emerged from conversations at the University of Aarhus in October 2014—in the AURA program (Aarhus University Research on the Anthropocene)—where participants collectively generated the concept for the traumatic changes seen in human-tended farms, pastures, forests, and finally, enclosed plantations predicated on private property and reliant on slave labor and other forms of exploited, alienated, and usually spatially transported labor. See "Publications,"

AURA: Aarhus University Research on the Anthropocene, http://anthropocene.au.dk /publications/, updated October 26, 2010.

52. "Capitalocene" is a term attributed to Andreas Malm (2015) and Jason W. Moore (2016, 2017), who locate the rise of capitalist society in the year 1450, corresponding with the European formation of capitalism. This dating also places the Capitalocene in historical parallel with Anthropocene theories that emphasize colonial expansion as fundamental to the epoch's formation. The designation Capitalocene is meant to dislodge the industrial revolution as the primary impetus for anthropocenic changes. However, it is also important to note that the industrial revolution initiates a new "means of production" (in a Marxist sense), which takes place within a capitalist "mode of production," and thus represents a specific form of capitalist accumulation. To eschew the importance of that late nineteenth-century moment (the rise of industrialism) and how it convened capitalism and the environment in very specific ways would be a mistake. In other words, the operations of capital and industrialism cannot, at this point in time, be analytically separate. However, I do agree with Moore, and with Isabelle Stengers (2015) as well, that Anthropocene discourse, and perhaps intervention, risks becoming neo-Malthusianism (often as depopulation rhetoric), too technophilic (as in, "we can engineer our way out of this"), and can become a set of tropes that overlook inequalities. Finally, while Capitalocene proponents find capitalism as the primary force driving toward ecological degradation, it is also true that we continue to live with emissions from the (former) noncapitalist world (e.g., the USSR and China under actually existing socialism).

53. Alternatively, the Anthropocene can be seen as crystallizing capitalism with nature. See Swyngedouw 2010.

54. I thank Kalyanakrishnan Sivaramakrishnan for the phrasing "velocities of change," which he proposed during our seminar in the Yale MacMillan Agrarian Studies program. See Steffen et al. 2015 on the Great Acceleration.

1. WIND

1. See Barad 2007 on intrarelations; Ingold 2007 on touching "in" wind.

2. Both "aeolian" and "eólica" draw their etymology from Aeolis. I want to signal that link and also underscore the linguistic relationship between the terms used in Mexico and "the aeolian" as a concept. Los eólicos is the Spanish term commonly used in Oaxaca to designate wind park developments (or the turbines themselves), and wind-generated electricity is energía eólica. Resistance to the proliferation of wind parks is commonly known as the antieólico struggle.

3. See the introduction to "Life above Earth" (Howe 2015a).

4. See Harvey and Knox 2015, 6–15, on how roads (or in this case, roads transformed into streets) are spaces of projection and material transformation where we can observe a negotiation between generic and specific forms of knowledge. Copaving by government and corporate entities in La Ventosa reflects a similar concentration of specialized knowledges and expert intervention. See also Dalakoglou and Harvey 2012; Masquelier 2002.

5. Large-scale energy projects, such as wind parks, are prone to follow a "developmentalist" model (Turner and Fajans-Turner 2006, 2) that is capital intensive, dependent upon both state and private financing, and oriented toward installing physical infrastructures. In my discussion of wind park development(s) throughout this book, I am building from several overlapping discussions in anthropology that take "development" as their central engagement. While I do not offer here a specific prognosis on development writ large, I do advance the proposition that pursuing the development trajectories of carbon energy acquisition and distribution cannot suffice in the present. For more on development (and "underdevelopment"), see, for example, Crewe and Axelby 2013; Edelman and Haugerud 2005; Escobar 1994; Ferguson 1990; Frank 1969; Kearney 1986; Li 2007.

6. See chapter 2 in *Energopolitics*, the companion volume to this one.

7. Don José does not describe himself as an "aeolian subject," but he is clear that his life has, in fact, been deeply contoured by wind's effects and powers.

8. Terán's poem was originally written in Zapotec and was translated by the author into Spanish. The Spanish-to-English translation was done by David Shook, and it appeared in English in the April 2009 edition of the internationally acclaimed *Poetry* magazine. Also see Terán 2009, 2015; Terán and Shook 2015.

9. The term *binnizá* (people of the clouds) is often used to mark Zapotec ethnicity in the isthmus. However, in our conversation, Terán used the term "Zapotec."

10. On Zapotec language, see, for example, Augsburger 2004. See OLAC (Open Language Archives) 2018 for a comprehensive list of scholarly work on isthmus Zapotec language from the 1940s to the present (including lexicons, grammar, literacy, etc.).

11. See Adey 2014, 15.

12. See Barad 2003. On "agential realist ontology" and "intra-acting 'agencies,'" see Barad 2007, 136–39. Also see Mol 2002 on the onto-specificities formed in medical practices and the social production of disease.

13. For Ingold (2007), the wind shows us that we cannot touch unless we first feel. Wind's relational force is also captured in his statement, "To feel the wind is to experience [a] commingling" (S29).

14. Irigaray's *The Forgetting of Air in Martin Heidegger* (1999) was a response to Heidegger's prioritization of Logos and earth in his formulation of Dasein. Irigaray contends that the omission of air is consequential. As she sees it, no philosophy of being can exist without a philosophy of breathing (315).

15. De Garay 1846, 35. De Garay's team was tasked with surveying the region for the purposes of a future transisthmus canal, and their exploration appears to have been contracted by the British.

16. The Institute of Electrical Studies and the National Water Commission (CONAGUA) was also involved in the wind-mapping project.

17. Elliott et al. 2003, 21.

18. For work on isthmus politics, see, for example, Binford 1985; Campbell 1990; Campbell et. al. 1993; Chassen-López 2004; Conant 2010; Kraemer Bayer 2008; Nader 1990; Rubin 1998; Stephen 2013; Warman 1993. On Mexico and indigeneity,

see Liffman 2014. On Mexico more generally, see Sánchez Prado 2015; Wolf and Hansen 1967.

19. In comparison, see a Texas utility company's offering of free nighttime electricity to customers due to the combination of nighttime wind power generation and lesser nighttime demand: Krauss and Cardwell 2015.

20. These debates are addressed in depth in chapters 3 and 5 of this volume.

21. De la Bellacasa 2011, 90.

22. For further reading on the politics of collectivity and proprietorship, see Ferry 2005.

23. See chapter 1 in *Energopolitics* for more detail on Sergio and the Ixtepec proposal.

24. The collective estate (bienes comunales) in Ixtepec was established in 1944 and covers 29,440 hectares (approximately 114 square miles) of land.

25. Also see more detail in *Energopolitics*, chapter 1.

26. One question that was posed in our survey of La Ventosa was, "To whom does the wind belong?" See *Energopolitics*, chapter 2, for more detail on the survey process and results. We spent approximately two weeks in La Ventosa, working with local residents and a handful of researchers from Juchitán to conduct a comprehensive survey of the community where every home was queried about the residents' feelings and experiences with wind park development.

27. See chapter 2 of this volume.

28. The term *clima* in Spanish designates "climate" in two senses, meteorological and political. It is also a term commonly used for "weather" (as is *tiempo*). On the changing climate, Terán noted, "He escuchado algo sobre los cambios de la naturaleza o el clima." In an interview, three young antieólico protestors observed, "El mundo está muy mal en la cuestión de clima ¿no?" Or, as Governor Cué enunciated at the FIER (Foro Internacional de Energía Renovables) symposium, "Porque el clima está cambiando, eso es exactamente lo que significa el cambio climático, el clima está cambiando."

29. For more on birds and other nonhuman life and ecological considerations, see chapter 5 of this volume.

30. How the winds have—at least potentially—been distorted by turbines is a question that remains. This was not the most common worry in the isthmus. More pressing political questions about land and bribery and intimidation were the most present concerns among istmo residents, as well as among officials in the state and national capitals seeking to manage the effects of the wind parks. But the ways that the winds had changed, would change, or might change were not inconsequential.

31. In his essay "Earth, Sky, Wind, and Weather" (2007), Tim Ingold sketches the qualities of an "open" world, where persons and things relate not as closed, separate, autonomous forms reacting to one another, but are instead constituted by their common immersion in a medium of generative flux. That medium, for Ingold, is air, wind, and weather. He posits, if earth and sky are viewed as separate but complementary hemispheres, furnished with "environment"—for example, trees, rocks, mountains— then we face a phenomenological dilemma: "If we are out in the open, how can we also be in the wind?" (S19).

1. It should be noted that this would have been the largest "single-phase" wind park in Latin America, meaning that it would be installed in one phase rather than iteratively.

2. I draw from a specifically anthropological interpretation of ethics that is collective and environmentally rooted. James Faubion (2011, 119) articulates this sensibility well when he writes, "Neither methodologically nor ontologically does an anthropology of ethics have its ground in the individual. The population of its interpretive universe is instead one of subjects in or passing through positions in environments. It is thus a population not of atomic units but of complex relata. Its subjects are for their part already highly complex. . . . Like the typical human being, the ethical subject, even when only an individual human being, is thus already always of intersubjective, social and cultural tissue. Its parts are never entirely its own. Ethical subject is not an abstraction." On ethics and climate change, also see Faubion 2011.

3. This ethos carried through to the level of international financial institutions such that the Interamerican Development Bank declared that "anything green" was an investment priority.

4. An ethical actor may, in addition to collective and grounded principles as noted above, nurture a reflexive ethics or "the kind of relationship you ought to have with yourself, rapport a soi" (Foucault 1997, 263). For Foucault, ethics is a determination of how individuals are meant to constitute themselves as a moral subject of their own actions.

5. Also see *Energopolitics*, chapter 2, for another iteration of this event.

6. The name Cymene (pronounced like "symmetry," but with "knee" replacing "tree") is not easily pronounceable in Spanish, and therefore many of our interlocutors referred to me as "Ximena" (pronounced "he-mena"), which has a similar pronunciation and is a familiar Spanish-language name.

7. The cohabitability of cattle and turbines was a common theme in our fieldwork; many farmers and ranchers were quick to point out how easily the two could coexist. Indeed, there are many indicative representations of this in online searches for wind parks in the isthmus, revealing cattle lazily meandering between the towers. Cattle ranchers with whom we spoke felt confident that their land was resilient enough for both. Some questions remained about water tables in the isthmus and whether their relatively shallow disposition would render it difficult to irrigate with massive concrete-and-rebar turbine bases inserted at points within them, but this was not a much-aired concern. More common was the lament that agriculture and cattle ranching were not being taken up by younger generations of istmeños who preferred instead to migrate north to become educated for white-collar careers or who found their way to cities across the country where more lucrative and perhaps less physically challenging and environmentally dependent work could be found.

8. See Mimiaga Sosa's publications on the wind sector of the isthmus, addressed primarily to policy makers and investors: Borja Díaz, Jaramillo Salgado, and Mimiaga Sosa 2005; Mimiaga Sosa 2009. On wind "speculation," see Galbraith and Price 2013.

9. See the potential illegality of the Ixtepec Potencia substation in chapter 1 of *Energopolitics*; Comisión Reguladora de Energía 2012.

10. See de Ita 2003.

11. On ejidos, bienes comunales, and land dispossession and distribution in Oaxaca, and in Mexico more generally, see Assies 2008; Benton 2011; Brown 2004; Castellanos 2010; Hoffmann 1998; Michel 2009; Zendejas 1995. Also see chapters 1 and 2 in *Energopolitics*.

12. See Brown 2004, 4.

13. It should be noted that some comunidades agrarias parceled agricultural land to individuals within the community to be worked or farmed as "their" land. Even in the case when a comunero might not be actively farming the land, the expectation is that it is still someone's land, even as it is legally still communally held and is, in fact, owned by the Mexican state. In this sense, some tracts of land can be viewed as nominally "private," or they may be more openly accessed by members of the asamblea. I thank one of the anonymous reviewers of this text for surfacing this critical detail.

14. See Alatout and Schelly 2010.

15. *Chilango*, they explained, was a term that emerged "many years ago, when the Spanish ships came to Vera Cruz. When the criollos arrived in the hot areas, they turned red (like the Huachinango [fish, red snapper] that is white in the water and then turns red when caught). So 'Chilango' is the name for those people who live in Central Mexico." "Chilango" is also commonly used in reference to residents of the urban metropole, Mexico City.

16. Throughout this book, it will be clear that the Mareña Renovables project specifically, and wind power in the isthmus more generally, was meant as an infrastructural apparatus that would provide renewable power for corporate customers. In addition, of course, wind power was intended to help remediate climate change and ensure the role of the Mexican state in doing so. While wind power parks are clearly infrastructural projects, my focus here is not on their infrastructural capacities but instead on their potential to raise concerns, exact environmental worries, and potentially disrupt norms of livelihoods in the places where they are sited. A growing body of work, some of which I draw upon here, is more specifically focused on infrastructures in their material forms (Anand 2017; Barry 2013; Harvey 2010; Khan 2006; Lockrem 2016) as well as their histories and futures (Carse 2014; Rodgers and O'Neill 2012; Schwenkel 2013; Star 1999). Here, I have elected to follow Brian Larkin's reading of infrastructures as mediational or "enabling" devices. As Larkin (2013) writes, infrastructures are enabling devices, moving flows of goods, people, and ideas. In the case I examine most deeply, Mareña Renovables, the potential of the park's creation enabled or channeled affective qualities—of both aspiration and worry—rather than a physical product (electricity). The park was an infrastructure that was never built and thus remains a case of an infrastructure in the future subjunctive.

17. See Komives et al. 2009.

18. On transmission capacity, see Comisión Federal de Electricidad 2012.

19. Under the Mexican system, the Comisión Reguladora de Energía estimates the level of developer demand for transmission and invites companies to respond with specific project proposals. The regulator then allocates capacity accordingly and levies holding fees.

20. Bryon 2013, n.p.

21. The neoliberal condition provides the ground for wind park development in Oaxaca at both the local and federal levels. However, my focus here is not on neoliberalism per se but rather on the ways that such projects in their "green" form can be made to reproduce many of the deficits of carbon-based energy development. Anthropological work on neoliberalism in Latin America and elsewhere is vast, but for especially pertinent works, see, for example; Comaroff and Comaroff 2001; Edelman and Haugerud 2005; Fisher 2009; Gledhill 2007; Hale 2006; Harvey 2005; Hill 2001; Lomnitz 2008; Ochoa 2001; Richard 2009; Rochlin 1997; Sawyer 2001; Schwegler 2008. Silvia Federici (2012) provides a particularly apt definition of neoliberalism: "the extension of the commodity form into every corner of the social factory."

22. For a more recent pronouncement of energy development, see Melgar 2017; Comisión Federal de Electricidad 2012.

23. This language is from the application for Clean Development Mechanism (CDM) status that was submitted to the UNFCCC by Vientos del Istmo SA de CV (the entity holding the rights to the development project that would be ultimately purchased and managed by Mareña) for 2006–9.

24. The injunction that ultimately stalled the Mareña project put it this way: "In the beginning of 2004 several wind companies in coordination with the federal government and state government of Oaxaca geographically distributed the territory of the Isthmus of Tehuantepec according to the quality of the wind, as indicated in the Atlas of Wind Resources (developed by NREL)."

25. In our conversations in Mexico City, a banking expert, who insisted on full anonymity, explained that "before the market collapsed and they became essentially worthless, CERs could actually amount to perhaps 7 percent of the total investment." This is equivalent to more than $30 million for a 200 megawatt park.

26. All of these quotes are from the Vientos del Istmo SA de CV CDM application, p. 11.

27. This provision came about with the 2008 LAERFTE reform (La Reforma Energética a la Ley de Aprovechamiento de Energías Renovables y el Financiamiento de la Transición Energética), which was intended to keep Mexico in compliance with its national programs for climate change mitigation, address technological concerns regarding transmission, and provision financial and legal security for investors. See Briones Gamboa 2008; SENER 2007; Tissot 2012; World Bank 2013.

28. Founded in 1996, Preneal was both a primary investor (or speculator) in the isthmus and a developer of wind, solarthermal, and biomass projects around the world.

29. The wind farm would have been Macquarie Mexican Infrastructure Fund's third investment in the country after a highway development project (2010) and cell tower assets (2011). According to Dow Jones, the two new majority owners, Tokyo-based Mitsubishi Corporation and Dutch pension fund manager PGGM, held a combined 67.5 percent stake in the Mareña Renovables park. See "Dutch Pension Fund and Mitsubishi Buy in to Macquarie's Mexican Wind Farm," Latin American Private Equity Venture Capital Association, February 24, 2012, http://lavca.org/2012/02/24/dutch -pension-fund-and-mitsubishi-buy- in-to-macquaries-mexican-wind-farm/.

30. According to an expert investor with many years of experience in the wind development sector in Mexico, despite the $89 million price tag paid by Macquarie, the accumulated project debt and the estimated funds that Preneal had invested into the project over a seven-to-eight-year period suggest that Preneal probably only walked away with about $1 million profit.

31. Also see *Energopolitics*, chapter 3, on FIER.

32. According to an Australian investment publication (affiliated with the *Wall Street Journal*), MMIF is Macquarie Group's first managed fund in Latin America and the first peso-denominated fund solely focused on Mexican infrastructure project investment. The fund focuses on investments in asset classes such as airports, water and wastewater, roads and rail, ports, and energy and utilities in addition to social and communications infrastructure. See Tan 2012.

33. Governor Cué clearly showed the government's support for the project when he said they would "accompany Mareña on the process of sensibilización."

34. Some of the figures that are noted are 1,000 pesos per hectare per year to the community of San Dionisio del Mar and Santa María; a projected 15–17 million pesos per year in electricity sales once the park is operational; and 1.5 million pesos per year to the ejidos of Zapata, Charis, and Álvaro Obregón for easement to three kilometers of roads.

35. For the isthmus region, see Torres Cantú 2016; Warman 1993; Villagómez Velázquez 2006.

36. The sandbar of Santa Teresa, where most of the park was to be sited, is held collectively as bienes comunales by the community of San Dionisio del Mar, whereas the land leading to the sandbar is held both in ejido and as private property.

37. See chapter 3 of this volume.

38. See "Diódoro y Jorge Castillo van a convencer a los huaves para instalar la eólica en San Dionisio," *Despertar*, October 1, 2012, http://www.despertardeoaxaca.com/?p =7297.

39. In an earlier report crafted by the municipal president of San Dionisio del Mar, installing a wind park had been part of the community development plan going forward.

40. Caciquismo and patronage is described in more detail in Bartra and Huerta 1978; Guerra 1992. On the influence of national political parties, especially the PRI and PRD in some isthmus communities, see chapter 2 in *Energopolitics*.

41. There is little scholarship on the ikojts population in the isthmus. However, see Diebold 1961.

42. In some comunidades agrarias, agricultural land is divided among individuals belonging to the ejido or comuna. While they do not directly own the land, they are taken to be the primary stewards of it and may acquire profits from said land. In the case of the Preneal contract (which then became the Mareña contract), the comuna of San Dionisio del Mar voted to agree to the park's construction, and the comisariado officials signed the actual contract on behalf of the asamblea/comuna membership. In this case, the comuneros themselves, as a whole, would receive the financial benefits of the contract; there was no stipulation that individualized parcels would affect the

payment structure. If the turbines were to have been placed on a farmer's individual parcel, one imagines the distribution of rental payments might have been affected since that farmer might have been prohibited from full use of his (and rarely, her) property. In the case of the Mareña park, no farmer's land was being affected, and no one's agricultural income would be threatened since the turbines were to be placed on a sandbar where no farming occurred. In addition to land tenure systems, as will be clear in coming chapters, there were significant political party tensions that came into play. Wind power companies in the early days, and perhaps even in the later stages, were undoubtedly challenged by the complex set of actors and interfaces at work when attempting to contract (in the logics of private property) that which was collectively managed if not individually owned.

43. Similar dynamics often pertain in other contexts as well. See, for example, Franquesa 2018 (Spain); Nadaï 2007 (France); Krauss 2010 (Germany); Pasqualetti 2011b (Scotland and California). A more seamless development of wind power, with the infrastructure notably owned by local residents, can be found in Kolbert 2008, "The Island in the Wind."

3. TRUCKS

1. Urban and regional planners have been attentive to the infrastructural demands of personal vehicles and their requisite roads and controls, as have philosophers (famously in Latour's 1996 study of the Aramis personal transport system). Some economists have gone so far as to distinguish the car as a singular machine in that it produces nothing but social and environmental externalities such as emissions (see, for example, Porter 1999). However, anthropological accounts of cars, trucks, and their social dynamics have been somewhat sparse, with some exceptions (see, for example, Alvarez and Collier 1994; Bright 1998; Miller 2001; Lochlann Jain 2004; Bohren 2009; Lutz 2014), in addition to cases based on other motorized transportation forms of such as motorbikes, mototaxis, and microbuses (e.g., Moodie 2006). Since much of humanity's relationship with the world has become increasingly mediated by passenger vehicles over the last century, anthropological inattention to these vehicles is striking, particularly since this is the same time period in which anthropology has grown and matured. It is a lacuna of a kind when juxtaposed against the many pages that the discipline has devoted to other material entities such as food, artisanal objects, or clothing. Even when modernist projects have become central to anthropological analysis—as in the studies of finance, scientific knowledge production, or the military to name a few—vehicles have still remained largely in the background, existing as a tool of modernity rather than as a locus of modernist practice. The advent of the Anthropocene, however, would seem to lead us toward the multiple ways that vehicles matter.

2. In a discussion of how nature has been taken as articulate, communicative, and intentional (here, primarily in reference to Latour), Kirby describes "passages of metamorphosis where the communication between matter and form is mutually enabled," suggesting that matter is both of and exceeding nature (2008, 227).

3. "Car culture," as Catherine Lutz (2014) has written, can be taken as emblematic of twentieth-century aspirations associated with individualism, speed, and achievement. By midcentury, the number of miles that private vehicles traveled in the US increased dramatically across the country. With the advent of suburbanization in the post–World War II period, the era correlated with the Great Acceleration.

4. One cannot really think of trucks without thinking of oil and its infrastructures. As the lifeblood of most cars and trucks occupying the world today, oil has also proven to be an important analytic substance for viewing the machinations of power, inequality, and regimes of knowledge that have structured the carbon age. See, for example Appel, Mason, and Watts 2015; Barry 2013, 2015; Guyer 2015; Appel 2012; LeMenager 2014; Mitchell 2011; Rogers 2015; Watts 2015.

5. See Alaimo and Hekman 2008 on the consequentialness of certain things over others in shifting contexts.

6. Donald Winnicott (1953, 1964) is one of the best known.

7. See Descola 2013a on the role of cognitive science in the development of his four ontologies of human/nature relating.

8. Like regimes of value, in which the commodity form is deemed a product of contextual social factors rather than a definite stage of economic development, varying regimes of modernity might appear coherent within particular cultural and political-economic contexts. At the same time, this coherence "may be highly variable from situation to situation" (Appadurai 1986, 15).

9. Section 22 is a Oaxacan teachers' union that is renowned for its ability to mount significant resistance; members of "el 22" were also part of APPO, which was able to occupy central avenues in the state capital, and which would, in turn, face persecution by state authorities. See, for example, Hernández Navarro 2006; Howell 2009; Stephen 2013.

10. It was unclear whether he meant to mock bureaucratic forms or mimic communist rhetoric, but in either case, the forces had been gathered.

11. On Charis and revolutionary impacts in the isthmus region, see de la Cruz 1993.

12. See chapter 4 of this volume for further details on this incident.

13. For more on this event, see chapters 4 and 6 of this volume.

14. See also Lomnitz 2005.

15. The claim that Bettina Cruz's neighbors did not know her whereabouts strains credulity given that residents of Juchitán are generally well aware of their neighbors' comings and goings. It is not impossible that they were unaware, but the more likely case is that this was an act of refusal to disclose any information to a menacing stranger in the night.

16. This is from a Facebook post by David Henestrosa, a reporter working in the region, who carefully documented wind power controversies.

17. See chapter 4 of this volume for more on the agreement and its contingencies.

18. Barry 2013; Holbraad 2007.

19. This is not unlike Heiddeger's "tool being."

1. What they procured was an anemometer, the technical term for the device that measures the speed of wind in a given location.

2. I thank one of the anonymous manuscript reviewers for the insight that the adoption of the title *inconforme* carries within it, in the context of Mexican politics, a powerful sense of "refusal" (see also Simpson 2016), whereby consensus and silent dissent are implicitly critiqued by the very claim to nonparticipation that "inconformity" demands.

3. On political struggle in the isthmus and beyond, see Arrioja Díaz Viruell and Sánches Silva 2012; Bailón Corres and Zermeño 1987; Campbell et al. 1993; Chassen-López 2004; Clarke 2000; Poole 2007; Tutino 1980.

4. On binnizá (Zapotec) practices, political resistance, and sovereignty, see Campbell 1990; Chiñas 1975; de la Cruz 2007; King 2012; Kraemer Bayer 2008; Münch Galindo 2006; Rojinsky 2008; Royce 1974; Whitecotton 1985. On Oaxaca more generally, see, for example, Chibnik 2003; Cohen 2004; Stephen 2005.

5. See chapter 2 in *Energopolitics*.

6. See Rubin 1998.

7. As further evidence of what he called "context," he continued, "In the isthmus they have these incredible natural resources and cultural riches, but they are enormously prone to conflict . . . all the way back to the conquista."

8. See, for example, Conant 2010; Graeber 2002.

9. Critics of the resistance regularly referenced (in press releases, for instance) that UCIZONI's longtime leader, Carlos Beas, was born in Chile and that Rodrigo was not from the isthmus originally but from another state in Mexico.

10. On the movement #YoSoy132, see Bacallao-Pino 2016.

11. For example, Carlos, a fervent activist and victim of violence linked to wind development, described COCEI in the following way: "They've converted from a socialist movement into a fascist bloc today. One hundred percent corrupt." Of the six factions currently in existence, he added, "none of these factions have the interests of the people in mind."

12. See Gudynas 2009; Howe 2014; Oceransky 2009; Turner and Fajans-Turner 2006.

13. See chapter 3 of this volume.

14. He was later accused repeatedly of having absconded with a portion of the proceeds in the amount of 6 million pesos; stories followed about posh homes he had purchased in resort communities.

15. See *Energopolitics*, chapter 2, on the role of national parties in the isthmus.

16. A February 9, 2103, article in *El Despertar* by Rebeca Luna Jiménez and Neomí López Cristóbal pithily described the importance of Álvaro Obregón in the conflict: "The greater part of the Barra de Santa Teresa belongs to the bienes comunales of San Dionisio del Mar, but the entrance to the barra is located in Álvaro Obregón. Before the Huaves refused to agree to the passage of trucks and construction materials across their territory, Mareña Renovables—and its antecedent, Preneal—corrupted the Zapotec municipal authorities of Álvaro Obregón in order to place a conduction cable

and transport materials to construct the turbines and docks. The authorities [comisariados] signed this agreement without consent from their respective assemblies who now block the entrance to the barra."

17. For an account of what constitutes the legal expectations associated with free, prior, and informed consent, see, for example, Portalewska 2012.

18. From the Declaración: San Dionisio del Mar, September 18, 2012, available at Stop Corporate Impunity, https://www.stopcorporateimpunity.org/declaracion-de-san-dionisio-del-mar/?lang=es.

19. The reason for protesting in front of the Danish Embassy was to critique the Danish company, Vestas, for its role in the turnkey contract and construction material provided for the Mareña park. Protestors also attempted to convene an action in front of the FEMSA/Coca-Cola compound but were turned away by guards. See chapter 3 of this volume for further details on the Mexico City protest.

20. See Rosa Rojas, "Protestan frente a BID, Mitsubishi y Coca-Cola por eólicos de Tehuantepec," *La Jornada*, October 17, 2012.

21. See "Indigenous Groups Protest Mexico's Biggest Wind-Energy Project," Dow Jones Newswires, Fox Business online, October 17, 2012, accessed October 18, 2012.

22. See chapter 3 of this volume.

23. The PRD delegate from Ixtepec did later demand publicly that CFE allow the community to bid on access to the substation in Ixtepec. See chapter 1 in *Energopolitics*.

24. See coverage by Enrique Méndez and Rodrigo Garduño advocating that the Ixtepec comuna be allowed to bid. "Proponen plan eólico alterno para Oaxaca," *La Jornada*, October 19, 2012, http://www.jornada.unam.mx/2012/10/19/estados/037n1est. Also see chapter 1 in *Energopolitics*.

25. On February 9, 2013, *El Despertar* published the original 2006 contract for the Santa María portion of the Mareña Renovables park and ran a story (Rebeca Luna Jimenez, "Hasta la COCEI se vendió con Mareña Renovables") decrying the company's apparent payoffs to Héctor Sánchez López and Leopoldo de Gyves de la Cruz. The article notes, too, that in assembly meetings in 2006 and 2007, Preneal's proposition was unanimously rejected.

26. See chapter 3 of this volume.

27. Apparently the contractors were demanding identification cards from fisherfolk, and this is part of what sparked a renewed reaction.

28. It is worth restating here how unique the barra is as a particular form of land. The barra, because it is sand and thus not agricultural land, has never been farmed but has been historically a place from which one could set out to fish. While the wind park would have had no impact on agriculture (in this site at least), it threatened to disturb the waters surrounding the barra. Importantly, the quasi privatization of the space that the company's plans might have entailed was taken as a grave threat to fishing in the entire region because it would prevent access to the shoreline and fishing put-ins. This is largely why, as I will argue in subsequent chapters, fish came to spell the ultimate demise of the Mareña project.

29. Edith Avila was Eda's negotiating companion. On Edith and Eda and on Garza, see chapter 2 of this volume.

30. See more on the outcome in chapter 3 of this volume.

31. For more on confrontations with state police and death threats against spokespersons of the resistance, see chapter 1 of this volume.

32. Several months later, revisions were made to the general national amparo legislation, expanding its scope and protections of human rights. In our interview with Jorge Negrete, an attorney specializing in human rights, he explained that human rights law in Mexico was, at that time, "in a state of profound flux." In December 2011 a constitutional reform in the country fully recognized human rights for the first time; up until that time, the state had offered "guarantees" of protections, but as Negrete pointed out, "the guarantees are granted by the state and can be taken back." On April 1, 2013, the first reform (since 1936) of the amparo law created new possibilities for human rights provisions. Rights were there defined by international treaties to which Mexico was a signatory, and the reform effectively expanded which parties could plead for amparo protection. Shielding by amparo was no longer restricted to directly involved parties but was available to anyone with a "legitimate interest" in a concern. Peña Nieto's passing of the law garnered many responses. See, for example, "México y su Nueva Ley de Amparo," La Jornada, Jalisco (Edición Impresa), April 8, 2013. And see, "La Nueva Ley de Amparo protege los derechos humanos," CNN Mexico, February 14, 2013. The president himself described the change as "a modernizing reform that will enhance coexistence between individuals and the state."

33. An ejidal or communal agrarian amparo configures agrarian communities as socioeconomic and legal entities, as are its members (ejidatarios or comuneros). This type of amparo has specific procedures that go beyond general administrative concerns.

34. There were two pertinent amparos. The legal complaint filed with the magistrate in the Tribunal Unitario Agrario in Tuxtepec (Oaxaca) questioned the validity of the original meeting of the San Dionisio assembly that had approved the initial contract; its proposition was to annul the "acta," claiming it was flawed because the meeting(s) did not conform to agrarian law, and hence the contract itself was null and void. The amparo filed in Salina Cruz challenged the acts of authority by government entities (e.g., SEMARNAT) that granted permission for the Mareña park. Though tactically distinct, each of these judicial moves intended the same outcome: to stay the project. The result of working on both fronts was that the Salina Cruz amparo was the most expedient.

35. Following Convenio 169 of La Organización Internacional del Trabajo.

36. The following government offices were named in the amparo filed in Salina Cruz, critiquing a wide swath of permitting by state officials: Comisión Reguladora de Energía; Secretaría de Comunicaciones y Transportes; Dirección General de Puertos de la SCT; Delegación de la SCT en el Estado de Oaxaca; SEMARNAT; Dirección General de la Zona Federal Marítimo Terrestre y Ambientes Costeros de la SEMARNAT; Dirección General de Gestión Forestal y de Suelos de la SEMARNAT; Delegación Federal de la Secretaría del Medio Ambiente y Recursos Naturales en el Estado de Oaxaca; Dirección General de Impacto y Riesgo Ambiental de la Subsecretaría de

Gestión para la Protección Ambiental de la Secretaría del Medio Ambiente y Recursos Naturales en el Estado de Oaxaca; Comisión Nacional del Agua; Ayuntamiento Municipal de San Dionisio del Mar, Oaxaca.

5. SPECIES

1. In *Notebook B* in Darwin's Transmutation of Species series, he wrote, "I think," above his first evolutionary tree on page 36 in mid-July 1837.

2. A sensibility toward species fit well with the Enlightenment mandate to transform God's works into scientifically defined lives and processes. Early "great chain of being" debates in the emerging fields of the social sciences were driven by distinctions among species. This led to the insinuation that observable differences among animals could be transferred onto the phenotypes of human faces, resulting in racist depictions (see, for example, Nott, Glidden, and Patterson 1854; Morton 1844). "Species" also served as a keyword for anthropology's evolutionary warriors, both unilinear and multilinear. Since that time, species-inspired contortions marking human difference have been used in neo-Darwinian tracts, marring Darwin's original intent. See, for example, Hird 2009, esp. chap. 3.

3. Two points are worth noting about species, although there are several others that could also be made. First, the concept of species demands a rule of separation: "All beings, to maintain themselves, have turned others away from their own paths" (Latour 2013, 215). Second, Darwin's original formulation seems to have depended on double truths and paradoxes; one cannot say precisely when one species began and another ended (if it did). The history of life is messier, and thus it may be true, as Tim Morton writes, that the "punchline of Darwin's book is that there are no species and they have no origin" (Morton 2013, 29).

4. Thom van Dooren describes species in part as an evolutionary "achievement" that occurs over multiple lineages, places, and interspecies relationships (2014, 16).

5. Haraway 2008, 22.

6. See Sweetlove 2011.

7. In the first decade of the twenty-first century, three hundred new mammals were documented. For example, on new species revelations, see Conniff 2010.

8. See de Vos et al. 2014 for calculations resulting in the extinction estimate of 1,000–10,000 times the background rate. On the sixth mass extinction, also see Kolbert 2014; Klein 2014. On ethics, care, and disappearing species, see van Dooren 2014.

9. Myers and Knoll 2001, 5389.

10. Nixon 2013.

11. These particular species were selected from the expeditionary report of Don José de Garay (1846), who recorded sightings of these species in the Isthmus of Tehuantepec. What his report designates as an "amber tree" is commonly known as "sweetgum."

12. Bond and Bessire 2014, 442.

13. See Hartigan 2017.

14. For the Marx of this time, "Man" was gendered "men."

15. As a result of his more extensive powers and needs in contrast to other animals, Man, for Marx, benefits from having the most complex ties of all. This reveals itself in production, where objects that are not of immediate need are created. A greater range of things are made, more "beautiful" things are fashioned, and Man is able to reproduce the objects he finds in nature (Marx 1932, 75–76).

16. In his discussion of "species thinking" (2009), Chakrabarty is well aware of the risks of universalizing a singular humanity, transhistorically and transgeographically, especially in anthropocenic conditions, where some humans have been the source of far more carbon dioxide and other greenhouse gas emissions than the overwhelming majority of other humans. Residents of the global North (particularly since the advent of the steam engine) have burdened the current ecosystem to such an extent that Chakrabarty asks, "Why should one include the poor of the world—whose carbon footprint is small anyway—by use of such an all-inclusive term such as *species* or *mankind*?" He responds to his own question by following history more deeply than the modern threshold of a few hundred years, pointing, for example, to the advent of agriculture as a major earth-changing event of human initiative. Thus, he returns to the pragmatic value of species thinking in the present age. "Whatever our socioeconomic and technological choices, whatever the rights we wish to celebrate as our freedom, we cannot afford to destabilize conditions (such as the temperature zone in which the planet exists) that work like boundary parameters of human existence."

17. See Chakrabarty 2009; see also the introduction to this volume.

18. Reflections such as these have been on many people's minds of late, if in different terms. See, for example, Bennett 2010; de la Cadena 2015; Hartigan 2015; Helmreich 2009; Jasarevic 2015; Kohn 2013; Lien 2015; Paxson 2008; Raffles 2010; Tsing 2012, 2015.

19. The ability of nonhumans to direct human thought is an insight borrowed from Amitav Ghosh (2016). In a parallel form, Jane Bennett considers the ability of things—edibles, commodities, storms, metals—to impede or block human will and likewise see how they act as "quasi agents or forces with trajectories, propensities, or tendencies" (Bennett 2010, viii).

20. For Stengers, this is linked to the cosmopolitical proposal. The cosmos, she writes, "corresponds to no condition, establishes no requirements. It creates the question of possible nonhierarchical modes of coexistences among the ensemble of inventions of nonequivalence, among the diverging values and obligations through which the entangled existences that compose it are affirmed . . . thus [integrating] an ecology of practices [that involves multiple domains of living]" (2011, 356).

21. My use of "sphere" follows Sloterdijk 2014 in the sense that I intend the windsphere to be deglobalized and yet still function as a gathering force.

22. De Garay 1846, 67.

23. The IUCN was founded through the first director general of UNESCO in 1948 in order to establish a centralized agency to address environmental challenges. The organization has been globally recognized for its "red list" documenting endangered and threatened species.

24. De Garay's team was tasked with surveying the region for the purposes of a future transisthmus canal, and their exploration appears to have been contracted by the British.

25. De Garay 1846, 63–65.

26. Fabiana Li (2015) takes the notion of "equivalence" as an analytic to understand how conflicts regarding mining in Peru are conceptualized and how tensions nonetheless remain. For Li, equivalence is a practice of expertise (and technical mechanisms) that relies on quantification and comparison as well as a fraught political relationship and contestation regarding authoritative knowledge.

27. In a similar fashion, much of the expository material in this chapter is derived from reports and other official documents. Those elements of the chapter that comprise interviews or participant observation data were gathered in direct fieldwork.

28. Developers are responsible for reforestation of areas beyond what is removed for a project. SEMARNAT is also specific about species, noting that "you cannot just do the reforestation ad hoc, go plant a bunch of little pine trees in a deciduous forest. . . . Your proposal would be invalid in that case."

29. The logics of systems thinking are apparent here, evoking networks, actors, and interrelated actants, human and nonhuman. See, for example, Latour 2005; Law 2009; Law and Hassard 1999.

30. The possibility that the environmental system might improve rather than continue a dynamic of decline is not a consideration in Alberto's narrative.

31. See especially "Case Studies Part II," in United Nations Environment Programme (Division of Technology, Industry and Economics) "UNEP Studies of EIA Practice in Developing Countries," May 2003, http://www.unep.ch/etb/publications /Compendium.php.

32. The 2013 passage of federal environmental liability legislation (General Ecological Balance and Environmental Protection Law) may help to ensure closer adherence to environmental standards. See Llamas and González 2003.

33. See especially Mathews 2011.

34. Once the manifestación has been submitted to SEMARNAT, the project proposer has five days to publish his or her intentions in "a widely circulated press source in the region to be impacted," and for twenty days after the publication date of the announcement, public consultation is invited. Representatives of SEMARNAT go about gathering commentaries regarding potential repercussions as designated by the report. Importantly, this is also the time when possible impacts that are absent from the proposers' MIA can be challenged. During the consultation phase, SEMARNAT administrators can query other state officials and other experts. Scientists, university professors, NGOs, and private institutes can be tasked with providing opinions, feedback, or consultation. Following the public meeting, SEMARNAT has two to three months to receive further comments and to invite review from state-level environmental consultants, universities, biologists, and various federal commissions.

35. If a development project like a wind park negatively impacts a protected species, for example, it is automatically rejected by SEMARNAT.

36. Mareña did have to try again. SEMARNAT demanded that the company submit additional studies, including a yearlong monitoring of birds and bats as well as baseline studies and conservation plans—for marine turtles, the liebre Tehuana, and the cinnamon-tailed sparrow (*Aimophila sumichrasti*)—all of which were delivered to SEMARNAT between March and June 2011.

37. See Hecht, Morrison, and Padoch 2014; Mathews 2011.

38. Humans are thus made into a particular kind of species. If "culture" has long been used to distinguish humans from their animal others, here it is taken merely as a quality of species behavior. The medio físico report suggests that human populations bear an equivalence to animal species and thus, especially in the case of indigenous peoples, can be taken as a form of racist bureaucracy.

39. Yusoff 2013b, 208.

40. Andrew Pickering's thoughts in *The Mangle of Practice* (1995) are pertinent here in regard to the "machinic." He argues that human practices are intertwined in the "mangle" of scientific/bureaucratic work; human and material agency are reciprocal, but one is not reducible to the other, nor are they interchangeable (15–17, 21–23). His concern with actor-network theory is that it attributes equal actancy to all in the network. This is a mistake. In a memorable passage he writes, "I find it hard to imagine any combination of naked human minds and bodies that could substitute for a telescope, let alone an electron microscope" (15). Here, I want to call attention to the bureaucratic mode of institutions as "machinic," recognizing that these must be integrated with the affective practice and attunements that staff at SEMARNAT and other institutions are encouraged to develop.

41. Thinking with species here shares a kinship with Tim Morton's (2010) notion of the "mesh," an indivisibility not only between nature and humans but between and among nature, beings, and all other objects, forces, and matter. Following a Heideggerian, object-oriented ontology (and in conversation with Graham Harman), Morton views all life and nonlife as coexistent and conjoined. For distinct interpretations of the life/nonlife nexus, see Povinelli 2016 on geontologies, where she rejects the life/nonlife binary in social theory (and ethnographically); see Barad 2007 on quantum matter and the physics of entanglement.

42. In total, 145 bird species have been classified as extinct since 1500, including five species that have gone extinct in the wild but that retain populations in captivity. Other bird species currently categorized as "critically endangered" have likely gone extinct too, but this has not been verified. A total of seventeen such species are categorized as "critically endangered (possibly extinct)" and one as "critically endangered (possibly extinct in the wild)." Thus, a total of 163 bird species may have been lost in the last five hundred years. See "We Have Lost Over 150 Bird Species since 1500," Birdlife International, http://www.birdlife.org/datazone/sowb/casestudy/102, last updated 2017.

43. An endemic bird area (EBA) is defined as an area that encompasses the overlapping breeding ranges of two or more restricted-range land birds, such that the complete ranges of at least two species fall entirely within the boundary of the EBA. Following this definition, a total of 218 EBAs have been identified globally, covering

the ranges of 93 percent of restricted-range birds (2,451 species). The majority of EBAS (77 percent) are in the tropics and subtropics (Stattersfield et al. 1998). Also see Birdlife International, http://www.birdlife.org/eba, accessed June 3, 2015.

44. Globally, according to the IUCN Red List, the cinnamon-tailed sparrow is considered "near threatened."

45. One interpretation of elegy is that it symbolically kills the person or being that is mourned through naming, interpolation, or calling out. See, for example, Fuss 2013, which traces the elegiac impulses in poetic forms as a way to recover the speech of the dead and to use their citational existence as a form of ethical practice. I thank Tim Morton for noting the correspondence between elegy and the mode of listing species and their (potential) extinctions.

46. Bats live everywhere with the exception of the northern and southern circumpolar regions and a few remote islands; they are also threatened by habitat loss. See IUCN SSC Bat Specialist Group, http://www.iucnbsg.org.

47. Interestingly, this is the same language used by Mareña Renovables to describe the wind of the isthmus: a "treasure," a "resource." See chapter 6 of this volume.

48. See Medellín and Gaona 1999.

49. The IDB report claims that bats fly below the collision risk zone, between three and ten meters, not in the risk zone of between 40 and 120 meters.

50. Deloria 2006.

51. See Keck and Lakoff 2013.

52. Contaminative effects and their resulting associations show up both in the bodies of humans (Agard-Jones forthcoming; Fortun 2001, 2014; Petryna 2002) and in contaminative externalities (Sawyer 2004; Cepek 2012).

53. Species classify the vital difference between survival and extinction, but the sorting of species also indicates human prioritizations of life-kinds. Human taste is involved in decisions about which sorts of creatures or vegetation are to be protected and preserved. Many environmental movements, for instance, have worked for the preservation of large primates, other mammals, cute cuddlies, and "charismatic species," prioritizing them over soil-dwelling or deep-ocean-dwelling slimies.

54. Or one that can be killed but not sacrificed (Agamben 1998). See also Fassin and Pandolfi 2010; Jackson and Warren 2005.

55. Showing interest in turtle eggs, however, can also result in askance looks from vendors who seem to want to dare out-of-place foreigners to criticize their sale of a scandalous commodity, the product of an environmentally poignant creature represented by the papery-white spherical housing of turtles-never-to-be.

56. Crutzen and Stoermer 2000, 17.

57. See, for example, Descola 2013a, 2013b; Viveiros de Castro 1998.

6. WIND POWER, IN SUSPENSION

1. Construction of the park was scheduled to begin in March 2012 and to be completed by July 2013.

2. Octavio Velez Ascencio, "Advierte Gabino Cué: 'Amparo en caso eólico es mala señal para la inversión,'" *Las Noticias*, December 9, 2012.

3. This is a familiar refrain of primativity that appears to span the globe, often in the service of dispossession people from their ancestral homelands. See, for instance, West 2016.

4. Secretary Martinez knew full well that poor decisions had been made by both the current state government and the previous one. But with an eye toward preserving the authority of state functionaries currently in office, he also noted that the present governor (Gabino Cué) had many factions to serve. Whereas the previous governor (Ulises Ruis Ortiz, whose term was 2004–10) had only to placate one party, his own (the PRI), Cué, who was elected through a political party coalition, needed to juggle several sets of interests and concerns. From Martinez's perspective, the agents of state governance faced a more complex set of contingencies now than had been the case in earlier phases of wind power development. For his part, Governor Cué continued to exhort wind power industrialists, like AMDEE (Asociación Mexicana de Energía Eólica), to get behind the effort to stabilize Mareña, if only for the greater good of the future of wind power in the region.

5. In the online publication e-Oaxaca (February 1, 2013), for example, Rosa Nidia Villalobos González, the president of the PRI-dominated unit, the Permanent Commission of Renewable Energy Development, stated (as she had many times before) that the fact that the government would allow "groups headed by the Chilean, Carlos Beas and Rodrigo Peñalosa to rob Oaxaca of an investment of one billion dollars that would generate the jobs that our people demand, speaks very badly of a government that hasn't had the wisdom to apply the law." See http://www.e-oaxaca.mx/noticias /conflictos/15636--inadmisible-que-un-grupusculo-ahuyente-inversiones-asegura -rosa-nidia-villalobos.html. Accessed February 3, 2013.

6. Back at the barricade in Álvaro Obregón a few weeks later, we spoke with some of the older men from Zapata and Álvaro who had positioned themselves in the shade of the crumbling brick wall of the former hacienda. Catching up on the news, we inquired about the canal-clearing jobs, and they nodded in recognition. "Yes, some of our people from the resistance are doing that work now and getting paid too. Which is good." We asked, "So, does that mean that they have agreed to the Mareña park, the men who are doing the canal work?" Absolutely not. "They may work for the daily pay," one man assured us, "but at night, they sleep here, with us, in the hacienda, to protect the barricade."

7. See several commentaries posted in a video here, regarding fishing, sovereignty, and political corruption. "Gui Xhi Ro, Pueblo Libre, Álvaro Obregón, Oaxaca, Istmo de Tehuantepec," published November 26, 2014, https://www.youtube.com/watch?v =Go6tvTukblM.

8. By August 2013 the asamblea had voted to reestablish usos y costumbres and to deny candidates from the established political parties from running in municipal elections; it was at this time that some voting booths were burned.

9. See IDB website: http://www.iadb.org/en/civil-society/public-consultations /independent-consultation-and-investigation-mechanism-icim/public-consultation -on-the-proposed-independent-consultation-and-investigation-mechanism,5603 .html. "Document of the Independent Consultation and Investigation Mechanism" submitted November 20, 2014, section 1.2, p. 3, accessed November 28, 2014.

10. The Indian Law Resource Center provides legal assistance for indigenous people and was instrumental in helping to draft the United Nations Declaration on the Rights of Indigenous Peoples, which was adopted by the United Nations General Assembly in 2007.

11. Most comuneros in Mexico are campesinos (farmers), not fisherfolk. Consequentially, fisherfolk are less likely to obtain the legal protections afforded by bienes communales and ejidal membership. However, the stay that had been decreed by the judge in Salina Cruz had worked to secure local residents' customary use and land access rights. In other words, the judge's injunction was predicated upon the fact that people were being deprived of their right to access their collectively held and communally administered land, in this case, the Barra de Santa Teresa. Here, fisherfolk had prevailed even if they were not among those who filed the amparo.

12. Residents of the séptima neighborhood in Juchitán, where fishing is a major income source, noted that there were three or four fishing cooperatives in operation, with two hundred to three hundred fishermen in each cooperative.

13. For more on the findings, see Anaya 2015.

14. See ICIM/IDB, "Document of the Independent Consultation and Investigation Mechanism," November 2014, http://idbdocs.iadb.org/wsdocs/getdocument.aspx?docnum=39499203.

15. For more details on the IDB's relevant operational policies, see http://www.iadb.org/en/mici/relevant-operational-policies,8166.html, accessed November 28, 2014. The final report of September 2016 states that IDB management failed to comply with the bank's policies, including the environmental and safeguards compliance policy and the access to information policy among others. See http://indianlaw.org/mdb/-development-bank-confirms-mexico-wind-farm-project-violated-indigenous-peoples, accessed November 28, 2014.

16. See *Energopolitics*, chapter 5.

17. In mid-April 2013 the secretary general resigned his post, saying that he no longer had the confidence of the governor.

18. A YouTube video of the caravana through Juchitán to Álvaro Obregón shows hundreds marching. "Caravana en apoyo al pueblo de alvaro Obregon," February 17, 2013, https://www.youtube.com/watch?v=GJMGO-npwWQ. See especially the arrival of the caravana in Álvaro Obregón around minute 1:10.

19. In early January 2014, a spokesperson for the Dutch pension fund that had invested in the project announced, "It is dead." A Dutch newspaper later reported that the project had "moved to two other sites in the region." See Rosa Rojas, "Muerto, proyecto eólico en San Dionisio, Oaxaca: De Telegraaf," *La Journada*, January 9, 2014, http://www.jornada.unam.mx/2014/01/09/sociedad/035n1soc.

References

Abélès, Marc, ed. 2011. *Des anthropologues a l'OMC: Scènes de la gouvernance mondiale*. Paris: CNRS.

Adey, Peter. 2014. *Air: Nature and Culture*. London: Reaktion.

Agamben, Giorgio. 1998. *Homo Sacer: Sovereign Power and Bare Life*. Stanford, CA: Stanford University Press.

Agard-Jones, Vanessa. Forthcoming. *Body Burdens: Toxic Endurance and Decolonial Desire in the French Atlantic*. Durham, NC: Duke University Press.

Ahmed, Sara. 2014. "White Men." *Feminist Killjoys*, November 4. https://feministkilljoys.com/2014/11/04/white-men.

Aiello, José L., Jorge I. Valencia, Enrique Caldera Muñoz, and Vicente L. Gómez. 1983. *Atlas Eolico Preliminar de America Latina y el Caribe: Programa regional de energía eolica de OLADE*. America Central y el Caribe 2. Quito, EC: OLADE.

Alaimo, Stacy. 2010. *Bodily Natures: Science, Environment, and the Material Self*. Indianapolis: Indiana University Press.

Alaimo, Stacy. 2016. *Exposed: Environmental Politics and Pleasures in Posthuman Times*. Minneapolis: University of Minnesota Press.

Alaimo, Stacy, and Susan Hekman, eds. 2008. *Material Feminisms*. Bloomington: Indiana University Press.

Alatout, Samer, and Chelsea Schelly. 2010. "Rural Electrification as a 'Bioterritorial' Technology: Redefining Space, Citizenship, and Power during the New Deal." *Radical History Review* 107: 127–38.

Alexander, Catherine, and Joshua Reno, eds. 2012. *Economies of Recycling: The Global Transformation of Materials, Values and Social Relations*. London: Zed Books.

Alexander, Catherine, and Andrew Sanchez, eds. 2018. *Indeterminacy: Waste, Value and the Imagination*. New York: Berghan Books.

Alley, Richard. 2014. *The Two-Mile Time Machine: Ice Cores, Abrupt Climate Change, and Our Future*. Princeton, NJ: Princeton University Press.

Almeyra, Guillermo, and Rebeca Alfonso Romero. 2004. *El Plan Puebla Panamá en el Istmo de Tehuantepec*. Mexico City: Universidad de la Ciudad de México.

Alonso Serna, Lourdes. 2014. "La energía eólica y los espacios de poder en el Istmo de Tehuantepec." Paper presented at the Congreso Internacional de Pueblos Indios de América Latina, "Siglos XIX–XXI Avances, perspectivas y retos," Oaxaca.

Alvarez, Robert R., Jr., and George A. Collier. 1994. "The Long Haul in Mexican Trucking: Traversing the Borderlands of the North and the South." *American Ethnologist* 21, no. 3 (August): 606–27.

Anand, Nikhil. 2017. *Hydraulic City: Water and the Infrastructures of Citizenship in Mumbai*. Durham, NC: Duke University Press.

Anaya, S. James. 2015. "Observaciones del Profesor S. James Anaya sobre la consulta en el contexto del proyecto energía eólica del sur en Juchitán de Zaragoza." February 23. Consulta Indígena en Juchitán. Documentos Energía Eólica. https:// consultaindigenajuchitan.files.wordpress.com/2015/01/juchitan-observaciones -anaya.pdf.

Appadurai, Arjun. 1986. *The Social Life of Things: Commodities in Cultural Perspective*. Cambridge: Cambridge University Press.

Appel, Hannah. 2012. "Walls and White Elephants: Oil Extraction, Responsibility, and Infrastructural Violence in Equatorial Guinea." *Ethnography* 13 (4): 439–65.

Appel, Hannah, Arthur Mason, and Michael Watts, eds. 2015. *Subterranean Estates: Life Worlds of Oil and Gas*. Ithaca, NY: Cornell University Press.

Arrioja Díaz Viruell, Luis Alberto, and Carlos Sánches Silva, eds. 2012. *Conflictos por la tierra en Oaxaca: De las reformas borbónicas a la reforma agraria*. Michoacán, Mexico: El Colegio de Michoacán, UABJ.

Assies, Willem. 2008. "Land Tenure and Tenure Regimes in México: An Overview." *Journal of Agrarian Change* 8 (1): 33–63.

Augsburger, Deborah. 2004. "Language Socialization and Shift in an Isthmus Zapotec Community of Mexico." PhD diss., University of Pennsylvania.

Bacallao-Pino, Lázaro M. 2016. "Radical Political Communication and Social Media: The Case of the Mexican #YoSoy132." In *(R)evolutionizing Political Communication through Social Media*, edited by Tomaz Dezelan and Igor Vobic, 56–74. Hershey, PA: IGI Global.

Bailón Corres, Moisés J., and Sergio Zermeño. 1987. *Juchitán: Límites de una Experiencia Democrática*. México City: Instituto de Investigaciones Sociales.

Barad, Karen. 2003. "Posthumanist Performativity: Toward an Understanding of How Matter Comes to Matter." *Signs* 28 (3): 801–31.

Barad, Karen. 2007. *Meeting the Universe Halfway: Quantum Physics and the Entanglement of Matter*. Durham, NC: Duke University Press.

Barnes, Jessica. 2014. *Cultivating the Nile: The Everyday Politics of Water in Egypt*. Durham, NC: Duke University Press.

Barnes, Jessica, Michael Dove, Myanna Lahsen, Andrew Mathews, Pamela McElwee, Roderick McIntosh, Frances Moore, Jessica O'Reilly, Ben Orlove, Rajindra Puri, Harvey Weiss, and Karina Yager. 2013. "Contribution of Anthropology to the Study of Climate Change." *Nature Climate Change* 3: 541–44.

Barry, Andrew. 2013. *Material Politics: Disputes along the Pipeline*. West Sussex, UK: Wiley-Blackwell.

Barry, Andrew. 2015. "The Oil Archives." In *Subterranean Estates: Life Worlds of Oil and Gas*, edited by Hannah Appel, Arthur Mason, and Michael Watts, 95–107. Ithaca, NY: Cornell University Press.

Bartra, Roger, and Eugenia Huerta. 1978. *Caciquismo y Poder Político en el México Rural*. Mexico City: Siglo Veintiuno Editores.

Bebbington, Anthony. 2009. "The New Extraction: Rewriting the Political Ecology of the Andes?" NACLA *Report on the Americas* 42 (5): 12–20.

Behrends, Andrea, Stephen Reyna, and Guenther Schlee, eds. 2011. *Crude Domination: The Anthropology of Oil*. New York: Berghahn.

Bennett, Jane. 2010. *Vibrant Matter: A Political Ecology of Things*. Durham, NC: Duke University Press.

Benton, Allyson. 2011. "The Origins of Mexico's Municipal Usos y Costumbres Regimes: Supporting Local Political Participation or Local Authoritarian Control?" Centro de Investigación y Docencia Económicas. Documentos de Trabajo 226. February. http://libreriacide.com/librospdf/DTEP-226.pdf.

Binford, Leigh. 1985. "Political Conflict and Land Tenure in the Mexican Isthmus of Tehuantepec." *Journal of Latin American Studies* 17 (1): 179–200.

Bini, Elisabetta, and Giuliano Garavini, eds. 2016. *Oil Shock: The 1973 Crisis and Its Economic Legacy*. London: I. B. Tauris.

Bohren, Lenora. 2009. "Car Culture and Decision Making: Choice and Climate Change." In *Anthropology and Climate Change: From Encounters to Actions*, edited by Susan A. Crate and Mark Nuttall, 370–79. Walnut Creek, CA: Left Coast.

Bond, David, and Lucas Bessire. 2014. "Ontological Anthropology and the Deferral of Critique." *American Ethnologist* 41 (3): 440–56.

Bonneuil, Christophe, and Jean-Baptiste Fressoz. 2016. *The Shock of the Anthropocene: The Earth, History and Us*. London: Verso.

Booth, William. 2010. "Mexico Aims to Be a Leader in Emissions Reduction." *Washington Post*, November 29.

Borja Díaz, Marco Antonio R., Oscar A. Jaramillo Salgado, and Fernando Mimiaga Sosa. 2005. *Primer Documento del Proyecto Eoloeléctrico del Corredor Eólico del Istmo de Tehuantepec*. Mexico City: Instituto de Investigaciones Eléctricas.

Bowker, Geoffrey C., et al. 2010. "Toward Information Infrastructure Studies: Ways of Knowing in a Networked Environment." In *International Handbook of Internet Research*, edited by J. Hunsinger et al., 97–117. Amsterdam: Springer.

Boyer, Dominic. 2014. "Energopower: An Introduction." *Anthropological Quarterly* 87 (2): 309–34.

Boyer, Dominic, and George Marcus. Forthcoming. Introduction to *Collaborative Anthropology Today*. Ithaca, NY: Cornell University Press.

Braun, Bruce, and Sarah J. Whatmore, eds. 2010. *Political Matter: Technoscience, Democracy and Public Life*. Minneapolis: University of Minnesota Press.

Breglia, Lisa. 2013. *Living with Oil: Promises, Peaks, and Declines on Mexico's Gulf Coast*. Austin: University of Texas Press.

Bright, Brenda. 1998. "Heart Like a Car": Hispano/Chicano Culture in Northern New Mexico." *American Ethnologist* 25 (4): 583–609.

Briones Gamboa, Fernando. 2008. "Clima y vulnerabilidad social: conflictos políticos y repartición de riesgos en el Istmo de Tehuantepec (Oaxaca)." In *Aires y lluvias: Antropología del clima en México*, edited by Anna María Lammel, Marina Goloubinoff, and Esther Kratz, 615–38. Mexico City: Casa Chata/CIESAS.

Broome, John. 2012. *Climate Matters: Ethics in a Warming World*. London: Norton.

Brown, Jennifer. 2004. "Ejidos and Comunidades in Oaxaca, Mexico: Impact of the 1992 Reforms." In *Rural Development Institute Reports on Foreign Aid and Development*, 120. Seattle: Rural Development Institute.

Bryon, Justin. 2013. *Considerations of the Mexican Renewable Energy Market from an Investor Perspective [Capital Structuring]*. Brochure, May.

Caldera Muñoz, Enrique, et al. 1980. "Estudio preliminar y potencial de La Ventosa, Oax., para el aprovechamiento de la energía eólica." *Boletín IIE* 4 (8/9): 46–57.

Caldera Muñoz, Enrique, and R. Saldaña Flores. 1986. "Evaluación Preliminar del Potencial de Generación Eléctrica en la Zona de La Ventosa." Oaxaca, Instituto de Investigaciones Eléctricas. Technical report FE/01/14/2063/1-01/P.

Campbell, Howard. 1990. "Juchitán: The Politics of Cultural Revivalism in an Isthmus Zapotec Community." *Latin American Anthropology Review* 2 (2): 47–55.

Campbell, Howard, Leigh Binford, Miguel Bartolomé, and Alicia Barabas, eds. 1993. *Zapotec Struggles: Histories, Politics, and Representations from Juchitán, Oaxaca*. Washington, DC: Smithsonian Institution Press.

Candea, Matei. 2013. "Suspending Disbelief: Epoché in Animal Behavior Science." *American Anthropologist* 115 (3): 423–36.

Carse, Ashley. 2014. *Beyond the Big Ditch: Politics, Ecology, and Infrastructure at the Panama Canal*. Cambridge, MA: MIT Press.

Castellanos, M. Bianet. 2010. "Don Teo's Expulsion: Property Regimes, Moral Economies, and Ejido Reform." *Journal of Latin American and Caribbean Anthropology* 15 (1): 144–69.

Cepek, Michael L. 2012. "The Loss of Oil: Constituting Disaster in Amazonian Ecuador." *Journal of Latin American and Caribbean Anthropology* 17 (3): 393–412.

Chakrabarty, Dipesh. 2009. "The Climate of History: Four Theses." *Critical Inquiry* 35 (2): 197–221.

Chassen-López, Francie R. 2004. *From Liberal to Revolutionary Oaxaca: The View from the South, Mexico 1867–1911*. University Park: Penn State University Press.

Chen, Mel. 2012. *Animacies: Biopolitics, Racial Mattering and Queer Affect*. Durham, NC: Duke University Press.

Chibnik, Michael. 2003. *Crafting Tradition: The Making and Marketing of Oaxacan Wood Carvings*. Austin: University of Texas Press.

Chiñas, Beverly. 1975. *Mujeres de San Juan: La mujer zapoteca del Istmo en la economía*. Mexico City: Secretaría de Educación Pública (SepSetentas).

Choy, Timothy. 2011. *Ecologies of Comparison: An Ethnography of Endangerment in Hong Kong*. Durham, NC: Duke University Press.

Choy, Timothy, and Jerry Zee. 2015. "Condition: Suspension." *Cultural Anthropology* 30 (2): 210–23.

Clarke, Colin. 2000. *Class, Ethnicity, and Community in Southern Mexico: Oaxaca's Peasantries*. Oxford: Oxford University Press.

Clover, Joshua, and Juliana Spahr. 2014. *#Misanthropocene: Twenty-Four Theses*. Oakland, CA: Commune Editions.

Cohen, Jeffrey H. 2004. *The Culture of Migration in Southern Mexico*. Austin: University of Texas Press.

Comaroff, Jean, and John L. Comaroff, eds. 2001. *Millennial Capitalism and the Culture of Neoliberalism*. Durham, NC: Duke University Press.

Comisión Federal de Electricidad. 2012. *Programa de Obras e Inversiones del Sector Eléctrico 2012–2026*. Mexico City: Comisión Federal de Electricidad.

Comisión Reguladora de Energía. 2012. *Memoria Descriptiva: Temporadas Abiertas de Reserva de Capacidad de Transmisión y Transformación*. Mexico City: Comisión Reguladora de Energía.

Conant, Jeff. 2010. *A Poetics of Resistance: The Revolutionary Public Relations of the Zapatista Insurgency*. Oakland, CA: AK Press.

Conniff, Richard. 2010. "Meet the New Species." *Smithsonian*, August. https://www.smithsonianmag.com/40th-anniversary/meet-the-new-species-748819/.

Coole, Diana, and Samantha Frost, eds. 2010. *New Materialisms: Ontology, Agency, and Politics*. Durham, NC: Duke University.

Cornelius, Wayne A., and David Myhre, eds. 1998. *The Transformation of Rural Mexico: Reforming the Ejido Sector*. La Jolla: Center for US-Mexican Studies, University of California San Diego.

Coronil, Fernando. 1997. *The Magical State: Nature, Money, and Modernity in Venezuela*. Chicago: University of Chicago.

Crate, Susan A., and Mark Nuttall, eds. 2009. *Anthropology and Climate Change: From Encounters to Actions*. Walnut Creek, CA: Left Coast.

Crewe, Emma, and Richard Axelby. 2013. *Anthropology and Development: Culture, Morality, and Politics in a Globalised World*. Cambridge: Cambridge University Press.

Crutzen, Paul, and Eugene Stoermer. 2000. "The Anthropocene." *Global Change Newsletter* 41: 17–18.

Cruz Islas, Ignacio Cesar. 2013. "Energy Consumption of Mexican Households." *Journal of Energy and Development* 38 (1–2): 189–219.

Daggett, Cara. 2019. *The Birth of Energy*. Durham, NC: Duke University Press.

Dalakoglou, Dimitris, and Penny Harvey. 2012. "Roads and Anthropology: Ethnographic Perspectives on Space, Time and (Im)Mobility." *Mobilities* 7 (4): 459–65.

Davis, Mike. 2010. "Who Will Build the Ark?" *New Left Review* 61 (January/February): 29–46.

De Garay, José. 1846. *An Account of the Isthmus of Tehuantepec in the Republic of Mexico; with Proposals for Establishing and Communication between the Atlantic and Pacific Oceans, Based upon the Surveys and Reports of a Scientific Commission, Appointed by the Projector Don José de Garay*. London: J. D. Smith.

De Ita, A. 2003. *Mexico: The Impacts of Demarcation and Titling by PROCEDE on Agrarian Conflicts and Land Concentration*. Mexico City: Centro de Estudios para el Cambio en el Campo Mexicano, Land Research Action Network.

De la Bellacasa, Maria Puig. 2011. "Matters of Care in Technoscience: Assembling Neglected Things." *Social Studies of Science* 41 (1): 85–106.

De la Cadena, Marisol. 2015. *Earth Beings: Ecologies of Practice across Andean Worlds.* Durham, NC: Duke University Press.

De la Cruz, Víctor. 1993. *El general Charis y la pacificatión del México posrevoluciona-rio.* Mexico City: CIESAS.

De la Cruz, Víctor. 2007. *El pensamiento de los binnigula'sa: cosmovisión, religión y calendario con especial referencia a los binnizá.* Mexico City: CIESAS-INAH.

Deloria, Vine, Jr. 2006. *The World We Used to Live In: Remembering the Powers of the Medicine Men.* Golden, CO: Fulcrum.

Descola, Philippe. 2013a. *Beyond Nature and Culture.* Chicago: University of Chicago Press.

Descola, Philippe. 2013b. *The Ecology of Others.* Chicago: Prickly Paradigm.

De Vos, Jurriaan M., Lucas N. Joppa, John L. Gittleman, Patrick R. Stephens, and Stuart L. Pimm. 2014. "Estimating the Normal Background Rate of Species Extinction." *Conservation Biology* 29 (2): 452–62.

Diebold, A. Richard, Jr. 1961. "Bilingualism and Biculturalism in a Huave Community." PhD diss., Yale University.

Dietrich, Christopher R. W. 2008. "The Permanence of Power: The Energy Crisis, Sovereign Debt, and the Rise of American Neoliberal Diplomacy, 1967–1976." PhD diss., University of Texas at Austin.

Douglas, Mary. 1957. "Animals in Lele Religious Symbolism." *Africa: Journal of the International African Institute* 27 (1): 46–58.

Eakin, Hallie. 2006. *Weathering Risk in Rural Mexico: Climatic, Institutional, and Economic Change.* Tucson: University of Arizona Press.

Edelman, Marc, and Angélique Haugerud. 2005. "Introduction: The Anthropology of Development and Globalization." In *The Anthropology of Development and Globalization: From Classical Political Economy to Contemporary Neoliberalism*, edited by Marc Edelman and Angélique Haugerud, 1–75. Malden, MA: Blackwell.

Edwards, Paul. 2013. *A Vast Machine: Computer Models, Climate Data, and the Politics of Global Warming.* Cambridge, MA: MIT Press.

Eilperin, Juliet, and Brady Dennis. 2016. "US, Canada and Mexico Vow to Get Half Their Electricity from Clean Power by 2025." *Washington Post*, June 27. https://www.washingtonpost.com/news/energy-environment/wp/2016/06/27/u-s-canada-and-mexico-to-pledge-to-source-half-their-overall-electricity-with-clean-power-by-2025/?noredirect=on&utm_term=.b1c41eed0c9f.

Elliott, Dennis, Marc Schwartz, Steve Haymes, Donna Heimiller, and R. George. 2003. *Wind Energy Resource Atlas of Oaxaca.* Oak Ridge, TN: NREL, US Department of Energy.

Escobar, Arturo. 1994. *Encountering Development: The Making and Unmaking of the Third World.* Princeton, NJ: Princeton University Press.

Fabian, Johannes. (1983) 2002. *Time and the Other: How Anthropology Makes Its Object.* New York: Columbia University Press.

Fassin, Didier, and Mariella Pandolfi. 2010. *Contemporary States of Emergency: The Politics of Military and Humanitarian Intervention.* Cambridge, MA: Zone.

Faubion, James. 2011. *An Anthropology of Ethics*. Cambridge: Cambridge University Press.

Federici, Sylvia. 2012. "Feminism and the Politics of the Commons." In *The Wealth of the Commons: A World beyond Market and State*, edited by David Bollier and Silke Helfrich. Springfield, MA: Levellers. Available at: http://wealthofthecommons.org /essay/feminism-and-politics-commons.

Ferguson, James. 1990. *The Anti-Politics Machine: "Development," Depoliticization and Bureaucratic Power in Lesotho*. Minneapolis: University of Minnesota Press.

Ferry, Elizabeth Emma. 2005. *Not Ours Alone: Patrimony, Value and Collectivity in Contemporary Mexico*. New York: Columbia University Press.

Fisher, Edward F., ed. 2009. *Indigenous Peoples, Civil Society, and the Neo-Liberal State in Latin America*. Oxford: Berghahn.

Fortun, Kim. 2001. *Advocacy after Bhopal: Environmentalism, Disaster, New Global Orders*. Chicago: University of Chicago Press.

Fortun, Kim. 2014. "From Latour to Late Industrialism." *HAU: Journal of Ethnographic Theory* 4 (1): 309–29.

Foucault, Michel. 1997. "On the Genealogy of Ethics: An Overview of the Work in Progress." In *Ethics, Subjectivity and Truth*. Vol. 1 of *The Essential Works of Foucault, 1954–1984*, edited by Paul Rabinow, translated by Robert Hurley et al. New York: New Press.

Frank, Andre Gunder. 1969. *Latin America: Underdevelopment and Revolution*. New York: Monthly Review Press.

Franklin, Sarah. 2007. *Dolly Mixtures: The Remaking of Genealogy*. Durham, NC: Duke University Press.

Franquesa, Jaume. 2018. *Power Struggles: Dignity, Value and the Renewable Energy Frontier in Spain*. Bloomington: University of Indiana Press.

Fuss, Diana. 2013. *Dying Modern: A Meditation on Elegy*. Durham, NC: Duke University Press.

Gabrys, Jennifer. 2013. *Digital Rubbish: A Natural History of Electronics*. Ann Arbor: University of Michigan Digital Culture Books.

Galbraith, Kate, and Asher Price. 2013. *The Great Texas Wind Rush: How George Bush, Ann Richards, and a Bunch of Tinkerers Helped the Oil and Gas State Win the Race to Wind Power*. Austin: University of Texas Press.

Galeano, Eduardo. 1997. *Open Veins of Latin America: Five Centuries of the Pillage of a Continent*. New York: Monthly Review Press.

Gardiner, Stephen. 2011. *A Perfect Moral Storm: The Ethical Tragedy of Climate Change*. Oxford: Oxford University Press.

Ghosh, Amitav. 2016. "Ep. #40—Amitav Ghosh." Cultures of Energy Podcast, October 27. Center for Energy and Environmental Sciences. http://culturesofenergy.com /ep-40-amitav-ghosh1/.

Gledhill, John. 2007. "Neoliberalism." In *A Companion to the Anthropology of Politics*, edited by David Nugent and Joan Vincent, 332–48. London: Blackwell.

Graeber, David. 2002. "The New Anarchists." *New Left Review* 18: 61–73.

Grove, Jairus. 2016. "Response to Jedediah Purdy." In "Forum: The New Nature," *Boston Review*, January 11.

Gudynas, Eduardo. 2009. "Diez Tesis Urgentes Sobre El Nuevo Extractivismo: Contextos y Demandas Bajo el Progresismo." In *Extractivismo, Política y Sociedad*. Quito, EC: CAAP (Centro Andino de Acción Popular) and CLAES (Centro Latino Americano de Ecología Social).

Guerra, François-Xavier. 1992. "Los orígenes socio-culturales del caciquismo." *Anuario del IEHS* (7): 181–83.

Gupta, Akhil. 2015. "Suspension." Theorizing the Contemporary, *Cultural Anthropology* website, September 24. https://culanth.org/fieldsights/722-suspension.

Guyer, Jane I. 2015. "Oil Assemblages and the Production of Confusion: Price Fluctuations in Two West African Oil-Producing Economies." In *Subterranean Estates: Life Worlds of Oil and Gas,* edited by Hannah Appel, Arthur Mason, and Michael Watts, 227–46. Ithaca, NY: Cornell University Press.

Hale, Charles R. 2006. *Más que un Indio (More Than an Indian): Racial Ambivalence and Neoliberal Multiculturalism in Guatemala*. Santa Fe, NM: School of American Research Press.

Hallowell, A. Irving. 1926. "Bear Ceremonialism in the Northern World." *American Anthropologist* 28 (1): 2–163.

Haraway, Donna J. 1988. "Situated Knowledges: The Science Question in Feminism and the Privilege of Partial Perspective." *Feminist Studies* 14 (3): 575–99.

Haraway, Donna J. 1996. *SecondMillennium.FemaleMan©MeetsOncoMouseTM™: Feminism and Technoscience*. New York: Routledge.

Haraway, Donna J. 2008. *When Species Meet*. Minneapolis: University of Minnesota Press.

Haraway, Donna J. 2015. "Anthropocene, Capitalocene, Plantationocene, Chthulucene: Making Kin." *Environmental Humanities* 6: 159–65.

Hartigan, John. 2013. "Mexican Genomics and the Roots of Racial Thinking." *Cultural Anthropology* 28 (3): 372–95.

Hartigan, John. 2015. "How to Interview a Plant: Part 1." *Aesop's Anthropology*, November 17. http://www.aesopsanthropology.com/blog/?p=320. Accessed January 14, 2016.

Hartigan, John. 2017. *Care of the Species: Races of Corn and the Science of Plant Biodiversity*. Minneapolis: University of Minnesota Press.

Harvey, David. 2005. *A Brief History of Neoliberalism*. Oxford: Oxford University Press.

Harvey, Penelope. 2010. "Cementing Relations: The Materiality of Roads and Public Spaces in Provincial Peru." *Social Analysis* 54 (2): 28–46.

Harvey, Penny, and Hannah Knox. 2015. *Roads: An Anthropology of Infrastructure and Expertise*. Ithaca, NY: Cornell University Press.

Hecht, Gabrielle. 2014. *Being Nuclear: Africans and the Global Uranium Trade*. Cambridge, MA: MIT Press.

Hecht, Susanna B., Kathleen Morrison, and Christine Padoch, eds. 2014. *The Social Lives of Forests: Past, Present, and Future of Woodland Resurgence*. Chicago: University of Chicago.

Helmreich, Stefan. 2009. *Alien Ocean: Anthropological Voyages in Microbial Seas*. Berkeley: University of California Press.

Henning, Annette. 2005. "Climate Change and Energy Use: The Role for Anthropological Research." *Anthropology Today* 21 (3): 8–12.

Hernández Navarro, Luis. 2006. "Oaxaca: Sublevación y crisis de un sistema regional de dominio." CLACSO, Consejo Latinoamericano de Ciencias Sociales: Argentina. *OSAL, Observatorio Social de América Latina* 7 (20): 1515–3282.

Hetherington, Kregg. 2013. "Beans before the Law: Knowledge Practices, Responsibility, and the Paraguayan Soy Boom." *Cultural Anthropology* 28 (1): 65–85. https://doi .org/10.1111/j.1548-1360.2012.01173.x.

Hill, Sarah. 2001. "The Environmental Divide: Neoliberal Incommensurability at the US-Mexico Border." *Urban Anthropology* 30: 157–88.

Hird, Myra. 2009. *The Origins of Sociable Life: Evolution after Science Studies.* Basingstoke, UK: Palgrave.

Hoffmann, J. 2012. "The Social Power of Wind: The Role of Participation and Social Entrepreneurship in Overcoming Barriers for Community Wind Farm Development. Lessons from the Ixtepec Community Wind Farm Project in Mexico." MA thesis, Lund University Centre for Sustainability Studies.

Hoffmann, Odile. 1998. "Tierra, poder y territorio. El *ejido* como institución compleja." In *Dinámicas de la conformación regional*, edited by A. Alvarado, O. Hoffmann, J.-Y. Marchal, N. Minello y M. Pépin-Lehalleur, 53–92. México: CNRS-Colegio de México-ORSTOM.

Holbraad, Martin. 2007. "The Power of Powder: Multiplicity and Motion in the Divinatory Cosmology of Cuban Ifá (or Mana, Again)." In *Thinking through Things: Theorizing Artifacts Ethnographically*, edited by Amiria Henare, Martin Holbraad, and Sari Wastell, 189–225. New York: Routledge.

Howe, Cymene. 2014. "Anthropocenic Ecoauthority: The Winds of Oaxaca." *Anthropological Quarterly* 87 (2): 381–404.

Howe, Cymene. 2015a. "Life above Earth: An Introduction." Special section, "Openings and Retrospectives." *Cultural Anthropology* 30 (2): 203–9.

Howe, Cymene. 2015b. "Introduction: Energy, Transition and Climate Change in Latin America." Special section, "Energy, Transition and Climate Change in Latin America." *Journal of Latin American and Caribbean Anthropology* 20 (2): 231–41.

Howe, Cymene, and Dominic Boyer. 2015. "Aeolian Politics." Special issue, "Political Materials." *Distinktion: Scandinavian Journal of Social Theory.*

Howe, Cymene, and Dominic Boyer. 2016. "Aeolian Extractivism and Community Wind in Southern Mexico." *Public Culture* 28 (2): 215–35.

Howe, Cymene, Dominic Boyer, and Edith Barrera. 2015. "Los márgenes del Estado al viento: autonomía y desarrollo de energías renovables en el sur de México." Special section, "Energy, Transition and Climate Change in Latin America." *Journal of Latin American and Caribbean Anthropology* 20 (2): 285–307.

Howe, Cymene, Jessica Lockrem, Hannah Appel, Edward Hackett, Dominic Boyer, Randal Hall, Matthew Schneider-Mayerson, et al. 2015. "Paradoxical Infrastructures: Ruins, Retrofit, and Risk." *Science, Technology, and Human Values* 41 (3): 547–65. https://doi.org/10.1177/0162243915620017.

Howell, Jayne. 2009. "Vocation or Vacation? Perspectives on Movement That Drew International Interest to Oaxaca Teachers' Union Struggles in Southern Mexico." *Anthropology of Work Review* 3: 87–98.

Hughes, David. 2017. *Energy without Conscience: Oil, Climate Change, and Complicity.* Durham, NC: Duke University Press.

Hulme, Mike. 2011. "Reducing the Future to Climate: A Story of Climate Determinism and Reductionism." *Osiris* 26: 245–66.

Ingold, Tim. 2004. "Culture on the Ground: The World Perceived through the Feet." *Journal of Material Culture* 9 (3): 315–40.

Ingold, Tim. 2007. "Earth, Sky, Wind and Weather." *Journal of the Royal Anthropological Institute* 13 (April): S19–38.

Irigaray, Luce. 1999. *The Forgetting of Air in Martin Heidegger.* Austin: University of Texas Press.

Jackson, Jean E., and Kay B. Warren. 2005. "Indigenous Movements in Latin America, 1992–2004: Controversies, Ironies, New Directions." *Annual Review of Anthropology* 34: 549–73.

Jasanoff, Sheila. 2010. "A New Climate for Society." *Theory, Culture and Society* 27 (2–3): 233–53.

Jasarevic, Larisa. 2015. "The Thing in a Jar: Mushrooms and Ontological Speculations in Post-Yugoslavia." *Cultural Anthropology* 30 (1): 36–64.

Johnston, Barbara Rose, Susan E. Dawson, and Gary E. Madsen. 2007. "Uranium Mining and Milling: Navajo Experiences in the American Southwest." In *Indians and Energy*, edited by Sherry Smith and Brian Frehner, 97–116. Santa Fe: SAR Press.

Kearney, Michael. 1986. "From the Invisible Hand to the Visible Feet: Anthropological Studies of Migration and Development." *Annual Review of Anthropology* 15: 331–61.

Keck, Frederic, and Andrew Lakoff. 2013. "Figures of Warning," *Limn* 3 (June). https://limn.it/articles/figures-of-warning/.

Khan, Naveeda. 2006. "Flaws in the Flow: Roads and Their Modernity in Pakistan." *Social Text* 24 (4): 87–113.

King, Stacie M. 2012. "Hidden Transcripts, Contested Landscapes and Long-Term Indigenous History in Oaxaca, Mexico." In *Decolonizing Indigenous Histories*, edited by Maxine Oland, Siobhan M. Hart, and Liam Frink, 230–63. Tucson: University of Arizona Press.

Kirby, Vicki. 2008. "Natural Convers(at)ions: Or, What If Culture Was Really Nature All Along?" In *Material Feminisms*, edited by Stacy Alaimo and Susan Hekman, 214–36. Bloomington: Indiana University Press.

Kirby, Vicki. 2011. *Quantum Anthropologies: Life at Large.* Durham, NC: Duke University Press.

Kirksey, Eben. 2014. *The Multispecies Salon.* Durham, NC: Duke University Press.

Klein, Naomi. 2015. *This Changes Everything: Capitalism vs. the Climate.* New York: Simon and Schuster.

Klieman, Kairn. 2008. "Oil, Politics, and Development in the Formation of a State: The Congolese Petroleum Wars, 1963–68." *International Journal of African Historical Studies* 41 (2): 169–202.

Kohn, Eduardo. 2013. *How Forests Think: Toward an Anthropology beyond the Human.* Berkeley: University of California Press.

Kolbert, Elizabeth. 2008. "The Island in the Wind." *New Yorker*, July 7. https://www
.newyorker.com/magazine/2008/07/07/the-island-in-the-wind.

Kolbert, Elizabeth. 2014. *The Sixth Great Extinction: An Unnatural History*. New York: Picador.

Komives, Kristin, Todd M. Johnson, Jonathan D. Halpern, José Luis Aburto, and John R. Scott. 2009. "Residential Electricity Subsidies in Mexico: Exploring Options for Reform and for Enhancing the Impact on the Poor." World Bank Working Paper 160. Washington, DC: World Bank.

Kraemer Bayer, Gabriela. 2008. *Autonomía de los zapotecos del Istmo. Relaciones de poder y cultura política*. Mexico City: CONACYT/Plaza y Valdés.

Krauss, Clifford, and Diana Cardwell. 2015. "A Texas Utility Offers a Nighttime Special: Free Electricity." *New York Times*, November 8. http://www.nytimes.com/2015
/11/09/business/energy-environment/a-texas-utility-offers-a-nighttime-special-free
-electricity.html?_r=0.

Krauss, Werner. 2010. "The 'Dingpolitik' of Wind Energy in Northern German Landscapes: An Ethnographic Case Study." *Landscape Research* 35 (2): 195–208.

Lahsen, Myanna. 2005. "Seductive Simulations: Uncertainty Distribution around Climate Models." *Social Studies of Science* 35: 895–922.

Larkin, Brian. 2013. "The Politics and Poetics of Infrastructure." *Annual Review of Anthropology* 42: 327–43.

Latour, Bruno. 1996. *Aramis or the Love of Technology*. Cambridge, MA: Harvard University Press.

Latour, Bruno. 2004a. *Politics of Nature: How to Bring the Sciences into Democracy*. Cambridge, MA: Harvard University Press.

Latour, Bruno. 2004b. "Why Has Critique Run Out of Steam? From Matters of Fact to Matters of Concern." *Critical Inquiry* 30 (2): 225–48.

Latour, Bruno. 2005. *Reassembling the Social: An Introduction to Actor-Network Theory*. Oxford: Oxford University Press.

Latour, Bruno. 2013. *An Inquiry into Modes of Existence: An Anthropology of the Moderns*. Cambridge, MA: Harvard University Press.

Latour, Bruno. 2014. "Anthropology at the Time of the Anthropocene: A Personal View of What Is to Be Studied." Presentation delivered at the American Anthropological Association meeting.

Law, John. 2009. "Actor Network Theory and Material Semiotics." In *The New Blackwell Companion to Social Theory*, edited by Bryan S. Turner, 141–58. Malden, MA: Wiley-Blackwell.

Law, John, and John Hassard, eds. 1999. *Actor Network Theory and After*. Oxford: Blackwell, the Sociological Review.

LeMenager, Stephanie. 2014. *Living Oil: Petroleum in the American Century*. Oxford: Oxford University Press.

Li, Fabiana. 2015. *Unearthing Conflict: Corporate Mining, Activism, and Expertise in Peru*. Durham, NC: Duke University Press.

Li, Tania Murray. 2007. *The Will to Improve: Governmentality, Development, and the Practice of Politics*. Durham, NC: Duke University Press.

Lien, Marianne. 2015. *Becoming Salmon: Aquaculture and the Domestication of a Fish.* Berkeley: University of California Press.

Liffman, Paul. 2014. *Huichol Territory and the Mexican Nation: Indigenous Ritual, Land Conflict, and Sovereignty Claims.* Tucson: University of Arizona Press.

Liffman, Paul. 2017. "El agua de nuestros hermanos mayores: La cosmopolítica antiminera de los wixaritari y sus aliados." In *Mostrar y ocultar en el arte y en los rituales: Perspectivas comparativas,* edited by Johannes Neurath and Guilhem Olivier, 563–88. Mexico City: Universidad Nacional Autónoma de México-Instituto de Investigaciones Estéticas.

Llamas, Mauricio, and Antonio González. 2003. "Enactment of the Federal Law on Environmental Liability in Mexico." *Jones Day* (June). http://www.jonesday.com /enactment-of-the-federal-law-on-environmental-liability-in-mexico/.

Lochlann Jain, Sarah S. 2004. "'Dangerous Instrumentality': The Bystander as Subject in Automobility." *Cultural Anthropology* 19 (1): 61–94.

Lockrem, Jessica. 2016. "Bodies in Motion: Attending to Experience, Emotion, the Senses, and Subjectivity in Studies of Transportation." *Mobility in History* 7 (1): 50–57.

Lomnitz, Claudio. 2005. "Sobre reciprocidad negativa." *Revista de Antropología Social* 14: 311–39.

Lomnitz, Claudio. 2008. "Narrating the Neoliberal Moment: History, Journalism, Historicity." *Public Culture* 20 (1): 39–56.

Love, Thomas. 2008. "Anthropology and the Fossil Fuel Era." *Anthropology Today* 24 (2): 3–4.

Love, Thomas, and Anna Garwood. 2011. "Wind, Sun, and Water: Complexities of Alternative Energy Development in Rural Northern Peru." *Rural Society* 20: 294–307.

Lowe, Celia. 2010. "Viral Clouds: Becoming H5N1 in Indonesia." *Cultural Anthropology* 25 (4): 625–49.

Luna Jiménez, Rebeca, and Noemí López Cristóbal. 2013. "Mareña Renovables corrompió a todos." *Despertar,* February 9. http://despertardeoaxaca.com/marena -renovables-corrompio-a-todos/.

Lutz, Catherine. 2014. "The US Car Colossus and the Production of Inequality." *American Ethnologist* 41 (2): 232–45.

Lutz, Catherine, and Anne Lutz Fernandez. 2010. *Carjacked: The Culture of the Automobile and Its Effect on Our Lives.* New York: St. Martins.

Lynch, Barbara. 1982. *The Vicos Experiment: A Study of the Impacts of the Cornell-Peru Project in a Highland Community.* Washington, DC: USAID.

Malm, Andreas. 2015. "The Anthropocene Myth." *The Jacobin,* March 30. https://www .jacobinmag.com/2015/03/anthropocene-capitalism-climate-change/.

Marcus, George E. 2018. "Introduction: Collaborative Analytics." Theorizing the Contemporary, *Cultural Anthropology* website, July 27, 2017. https://culanth.org /fieldsights/1170-introduction-collaborative-analytics.

Margulis, Lynn. 1970. *Origin of Eukaryotic Cells.* New Haven, CT: Yale University Press.

Marx, Karl. 1932. *Economic and Philosophic Manuscripts of 1844.* Westminster, MD: Prometheus.

Mason, Arthur, and Maria Stoilkova. 2012. "Corporeality of Consultant Expertise in Arctic Natural Gas Development." *Journal of Northern Studies* 6 (2): 83–96.

Masquelier, Adeline. 2002. "Road Mythographies: Space, Mobility, and the Historical Imagination in Postcolonial Niger." *American Ethnologist* 29 (4): 829–56.

Massumi, Brian. 2009. "National Enterprise Emergency." *Theory, Culture, and Society* 26 (6): 153–85.

Mathews, Andrew 2011. *Instituting Nature: Authority, Expertise, and Power in Mexican Forests.* Cambridge, MA: MIT Press.

Matsutake Worlds Research Group. 2009. "A New Form of Collaboration in Cultural Anthropology: Matsutake Worlds." *American Ethnologist* 36 (2): 380–403.

Mauss, Marcel. 1979. *Seasonal Variations of the Eskimo: A Study in Social Morphology.* Translated by James J. Fox. London: Routledge.

McKibben, Bill. 1989. *The End of Nature.* New York: Random House.

McNeish, John-Andrew, and Owen Logan, eds. 2012. *Flammable Societies: Studies on the Socioeconomics of Oil and Gas.* London: Pluto Press.

Medellín, Rodrigo A., and Osiris Gaona. 1999. "Seed Dispersal by Bats and Birds in Forest and Disturbed Habitats of Chiapas, México." *Biotropica* 31 (3): 478–85.

Melgar, Lourdes. 2017. "3 Questions: Lourdes Melgar on Mexico's Energy Reform." *MIT Energy Initiative.* http://news.mit.edu/2017/3-questions-lourdes-melgar-mexico -energy-reform-0322.

Mentz, Steve. 2015. *Shipwreck Modernity: Ecologies of Globalization, 1550–1719.* Minneapolis: University of Minnesota Press.

Merleau-Ponty, Maurice. 2002. *Phenomenology of Perception.* New York: Routledge Classics.

Michel, Aurélia, 2009. "Los territorios de la reforma agraria: construcción y deconstrucción de una ciudadanía rural en las comunidades del Istmo oaxaqueño, 1934–1984." In *El Istmo mexicano: una región inasequible,* edited by E. Velázquez, E. Léonard, O. Hoffmann, and M.-F. Prévôt-Schapura, 455–99. Estado, poderes locales y dinámicas espaciales (siglos XVI–XXI). México: CIESAS-IRD.

Miller, Daniel. 2001. *Car Cultures.* London: Bloomsbury Academic.

Mimiaga Sosa, Fernando. 2009. "Corredor Eólico del Istmo de Tehuantepec." PowerPoint presentation. Secretaría de Economía del Gobierno del Estado de Oaxaca.

Mitchell, Timothy. 2011. *Carbon Democracy: Political Power in the Age of Oil.* New York: Verso.

Mol, Annemarie. 2002. *The Body Multiple: Ontology in Medical Practice.* Durham, NC: Duke University.

Monbiot, George. 2009. *Heat: How to Stop the Planet from Burning.* Boston: South End.

Moodie, Ellen. 2006. "Microbus Crashes and Coca-Cola Cash." *American Ethnologist* 33 (1): 63–80.

Moore, Jason W. 2016. *Anthropocene or Capitalocene?: Nature, History, and the Crisis of Capitalism.* New York: PM Press.

Moore, Jason W. 2017. "The Capitalocene, Part I: On the Nature & Origins of Our Ecological Crisis." *Journal of Peasant Studies* 44 (3): 594–630.

Morton, Samuel G. 1844. *An Inquiry into the Distinctive Characteristics of the Aboriginal Race of America and Catalogue of Skulls of Man.* Philadelphia: John Penington.

Morton, Timothy. 2010. *The Ecological Thought*. Cambridge, MA: Harvard University Press.

Morton, Timothy. 2013. *Hyperobjects: Philosophy and Ecology after the End of the World*. Minneapolis, MN: University of Minnesota Press.

Morton, Timothy. 2016. *Dark Ecology: For a Logic of Future Coexistence*. New York: Columbia University Press.

Münch Galindo, Guido, 2006. *La organización ceremonial de Tehuantepec y Juchitán*. Mexico City: UNAM-Instituto de Investigaciones Antropológicas.

Myers, Fred. 1988. "Burning the Truck and Holding the Country: Forms of Property, Time, and the Negotiation of Identity among Pintupi Aborigines." In *Hunter-Gatherers, II: Property, Power, and Ideology*, edited by T. Ingold, D. Riches, and J. Woodburn. London: Berg.

Myers, Natasha. 2016. "Photosynthesis." Theorizing the Contemporary, *Cultural Anthropology* website. January 21. https://culanth.org/fieldsights/790-photosynthesis.

Myers, Norman, and Andrew H. Knoll. 2001. "The Biotic Crisis and the Future of Evolution." *Proceedings of the National Academy of Sciences* 98 (10): 5389–92.

Nadaï, Alain. 2007. "Planning, Siting, and the Local Acceptance of Wind Power: Some Lessons from the French Case." *Energy Policy* 35: 2715–26.

Nadasdy, Paul. 2007. "The Gift in the Animal: The Ontology of Hunting and Human-Animal Sociality." *American Ethnologist* 34 (1): 25–43.

Nader, Laura. 1990. *Harmony Ideology: Justice and Control in a Zapotec Mountain Village*. Palo Alto, CA: Stanford University Press.

Nader, Laura. 2004. "The Harder Path—Shifting Gears." *Anthropological Quarterly* 77 (4): 771–91.

Nader, Laura, ed. 2010. *The Energy Reader*. Malden, MA: Wiley-Blackwell.

Nader, Laura, and Stephen Beckerman. 1978. "Energy as It Relates to the Quality and Style of Life." *Annual Review of Energy* 3: 1–28.

Nading, Alex. 2012. "Dengue Mosquitos Are Single Mothers: Biopolitics Meets Ecological Aesthetics in Nicaraguan Community Health Work." *Cultural Anthropology* 27 (4): 572–96. https://doi.org/10.1111/j.1548-1360.2012.01162.x.

Nixon, Rob. 2013. *Slow Violence and the Environmentalism of the Poor*. Cambridge, MA: Harvard University Press.

Nott, Josiah Clark, George R. Gliddon, and Henry S. Patterson. 1854. *Types of Mankind*. Philadelphia: Lippincott, Grambo.

Oceransky, Sergio. 2009. "Wind Conflicts in the Isthmus of Tehuantepec: The Role of Ownership and Decision-Making Models in Indigenous Resistance to Wind Projects in Southern Mexico." *The Commoner* 13 (winter): 203–22.

Ochoa, Enrique C. 2001. "Neoliberalism, Disorder, and Militarization in Mexico." *Latin American Perspectives* 28 (4): 148–59.

Oreskes, Naomi, and Erik Conway. 2011. *Merchants of Doubt: How a Handful of Scientists Obscured the Truth on Issues from Tobacco Smoke to Global Warming*. New York: Bloomsbury.

Ortner, Sherry. 1974. "Is Woman to Nature as Man Is to Culture?" In *Woman, Culture, and Society*, edited by Michelle Rosaldo and Louise Lamphere. Stanford, CA: Stanford University Press.

Pasqualetti, Martin J. 2011a. "Opposing Wind Energy Landscapes: A Search for Common Cause." *Annals of the Association of American Geographers* 101 (4): 1–11.

Pasqualetti, Martin J. 2011b. "Social Barriers to Renewable Energy Landscapes." *Geographical Review* 101 (2): 201–23.

Paxson, Heather. 2008. "Post-Pasteurian Cultures: The Microbiopolitics of Raw-Milk Cheese in the United States." *Cultural Anthropology* 23 (1): 15–47. https://doi.org/10.1111/j.1548-1360.2008.00002.x.

Petryna, Adriana. 2002. *Life Exposed: Biological Citizens after Chernobyl*. Princeton, NJ: Princeton University Press.

Pickering, Andrew. 1995. *The Mangle of Practice*. Chicago: University of Chicago Press.

Pinkus, Karen. 2016. *Fuel: A Speculative Dictionary*. Minneapolis: University of Minnesota Press.

Poole, Deborah. 2007. "Political Autonomy and Cultural Diversity in the Oaxaca Rebellion." *Anthropology Newsletter* (March): 10–11.

Portalewska, Agnes. 2012. "Free, Prior and Informed Consent: Protecting Indigenous Peoples' Rights to Self-Determination, Participation, and Decision Making." *Cultural Survival Quarterly* (December). https://www.culturalsurvival.org/publications/cultural-survival-quarterly/free-prior-and-informed-consent-protecting-indigenous.

Porter, Natalie. 2013. "Bird Flu Biopower: Strategies for Multispecies Coexistence in Việt Nam." *American Ethnologist* 40 (1): 132–48.

Porter, Richard C. 1999. *Economics at the Wheel: The Costs of Cars and Drivers*. Bingley, UK: Emerald Group.

Povinelli, Elizabeth. 2016. *Geontologies: A Requiem to Late Liberalism*. Durham, NC: Duke University Press.

Price, David H. 2016. *Cold War Anthropology: The CIA, the Pentagon, and the Growth of Dual Use Anthropology*. Durham, NC: Duke University Press.

Raffles, Hugh. 2010. *Insectopedia*. New York: Pantheon.

Rappaport, Roy. (1968) 2000. *Pigs for the Ancestors: Ritual in the Ecology of a New Guinea People*. Long Grove, IL: Waveland.

Rhoades, Robert, Xavier Zapata, and Jenny Aragundy. 2008. "Mama Cotacachi: Local Perceptions and Societal Implications of Climate Change, Glacier Retreat, and Water Availability." In *Darkening Peaks: Mountain Glacier Retreat in Social and Biological Contexts*, edited by Ben Orlove, Ellen Wiegandt, Brian H. Luckman, 218–27. Berkeley: University of California Press.

Richard, Analiese M. 2009. "Mediating Dilemmas: Local NGOs and Rural Development in Neoliberal Mexico." *Political and Legal Anthropology Review* 32 (2): 166–94.

Rochlin, James F. 1997. *Redefining Mexican "Security": Society, State, and Region under NAFTA*. Boulder, CO: Lynne Rienner.

Rodgers, Dennis, and Bruce O'Neill. 2012. "Infrastructural Violence: Introduction to the Special Issue." *Ethnography* 13 (4): 401–12.

Rogers, Doug. 2015. *The Depths of Russia: Oil, Power, and Culture after Socialism*. Ithaca, NY: Cornell University Press.

Rojinsky, David. 2008. "Manso de Contreras' Relación of the Tehuantepec Rebellion (1660–1661): Violence, Counter-Insurgency Prose, and the Frontiers of Colonial

Justice." In *Border Interrogations: Questioning Spanish Frontiers*, edited by Benita Sampedro Vizcaya and Simon Doubleday, 198–203. New York: Berghahn.

Roncoli, Carla, Todd Crane, and Ben Orlove. 2009. "Fielding Climate Change in Cultural Anthropology." In *Anthropology and Climate Change*, edited by Susan Crate and Mark Nuttall, 87–115. Walnut Creek, CA: Left Coast.

Royce, Anya Peterson. 1974. *Prestigio y Afiliación en una Comunidad Urbana: Juchitán, Oaxaca*. Austin: University of Texas Press.

Rubin, Jeffrey W. 1998. *Decentering the Regime: Ethnicity, Radicalism, and Democracy in Juchitán, Mexico*. Durham: Duke University Press.

Sánchez Casanova, Wendy Marilú. 2012. "Soberanía energética negociada: Dos versiones en el Istmo Oaxaqueño." Paper presented at the 2nd Congreso Nacional de Antropología Social y Etnología.

Sánchez Prado, Ignacio M. 2015. "Democracy, Rule of Law, a 'Loving Republic,' and the Impossibility of the Political in Mexico." Translated by Ariel Wind. *Política Común* 7.

Sawyer, Suzana. 2001. "Fictions of Sovereignty: Prosthetic Petro-Capitalism, Neoliberal States, and Phantom-Like Citizens in Ecuador." *Journal of Latin American Anthropology* 6 (1): 156–97.

Sawyer, Suzana. 2004. *Crude Chronicles: Indigenous Politics, Multinational Oil, and Neoliberalism in Ecuador*. Durham, NC: Duke University Press.

Scheer, Hermann. 2004. *The Solar Economy: Renewable Energy for a Sustainable Global Future*. London: Earthscan.

Schwegler, Tara A. 2008. "Take It from the Top (Down)? Rethinking Neoliberalism and Political Hierarchy in Mexico." *American Ethnologist* 35 (4): 682–700.

Schwenkel, Christina. 2013. "Post/Socialist Affect: Ruination and Reconstruction of the Nation in Urban Vietnam." *Cultural Anthropology* 28 (2): 252–77.

Scranton, Roy. 2015. *Learning to Die in the Anthropocene: Reflections on the End of a Civilization*. San Francisco: City Lights Books.

SENER. 2007. Energía eólica y la política energética mexicana. Ing. Alma Santa Rita Feregrino Subdirectora de Energía y Medio Ambiente, SENER. Monterrey, México, October.

Simpson, Audra. 2016. "Consent's Revenge." *Cultural Anthropology* 31 (3): 326–33.

Sloterdijk, Peter. 2014. *Globes: Spheres*. Vol. 2: *Macrospherology*. New York: Semiotext(e).

Sneath, David, ed. 2013. "Special Section: Climate Histories and Environmental Change: Evidence and its Interpretation." *Cambridge Journal of Anthropology* 31 (1): 51–155.

Star, Susan Leigh. 1999. "The Ethnography of Infrastructure." *American Behavioral Scientist* 43 (3): 377–91.

Stattersfield, Alison, Michael J. Crosby, Adrian J. Long, and David C. Wege. 1998. *Endemic Bird Areas of the World: Priorities for Biodiversity Conservation*. Cambridge, UK: BirdLife International.

Steffen, Will, Wendy Broadgate, Lisa Deutsch, Owen Gaffney, and Cornelia Ludwig. 2015. "The Trajectory of the Anthropocene: The Great Acceleration." *Anthropocene Review 2015* 2 (1): 81–98.

Steffen, Will, Paul J. Crutzen, and John R. McNeill. 2007. "The Anthropocene: Are Humans Now Overwhelming the Great Forces of Nature?" *Ambio* 36 (8): 614–21.

Stengers, Isabelle. 2010. "Including Nonhumans in Political Theory: Opening Pandora's Box?" In *Political Matter: Technoscience, Democracy, and Public Life*, edited by B. Braun and S. J. Whatmore, 3–35. Minneapolis: University of Minnesota Press.

Stengers, Isabelle. 2011. *Cosmopolitics I and II*. Minneapolis: University of Minnesota Press.

Stengers, Isabelle. 2015. *In Catastrophic Times: Resisting the Coming Barbarism*. Self-published, Open Humanities Press. http://openhumanitiespress.org/books /download/Stengers_2015_In-Catastrophic-Times.pdf.

Stephen, Lynn. 2005. *Zapotec Women: Gender, Class, and Ethnicity in Globalized Oaxaca*. 2nd ed. Durham, NC: Duke University Press.

Stephen, Lynn. 2013. *We Are the Face of Oaxaca: Testimony and Social Movements*. Durham, NC: Duke University Press.

Stewart, Kathleen. 2011. "Atmospheric Attunements." *Environment and Planning D: Society and Space* 29: 445–53.

Strathern, Marilyn. 1980. "No Nature, No Culture: The Hagen Case." In *Nature, Culture, and Gender*, edited by C. MacCormack and M. Strathern, 174–222. Cambridge: Cambridge University Press.

Strathern, Marilyn. 1992. *After Nature: English Kinship in the Late Twentieth Century*. Cambridge: Cambridge University Press.

Strauss, Sarah, Thomas Love, and Stephanie Rupp, eds. 2013. *Cultures of Energy*. Walnut Creek, CA: Left Coast.

Strauss, Sarah, and Ben Orlove, eds. 2003. *Weather, Climate, Culture*. Oxford: Berg.

Strong, Susan, Catherine Lutz, Jessica Katzenstein, and Amy Teller. n.d. "Gender and Citation in Cultural Anthropology." Unpublished manuscript.

Sweetlove, Lee. 2011. "Number of Species on Earth Tagged at 8.7 Million." *Nature*, August 23. http://www.nature.com/news/2011/110823/full/news.2011.498.html.

Swyngedouw, Erik. 2010. "Apocalypse Forever? Post-Political Populism and the Spectre of Climate Change." *Theory, Culture & Society* 27 (2–3): 213–32.

Tan, Gillian. 2012. "Macquarie Sells Mexican Wind Farm Stake." *Deal Journal Australia*, February 29. http://blogs.wsj.com/dealjournalaustralia/2012/02/29/macquarie -sells-mexican-wind-farm-stake/.

Terán, Víctor. 2009. "The North Wind Whips." *Poetry* (April). http://www .poetryfoundation.org/poetrymagazine/poem/185327.

Terán, Víctor. 2015. *The Spines of Love/Las Espinas del Amor/Ca Guichi Xtí Guendaranaxhii*. Translated by David Shook. New York: Restless Books.

Terán, Víctor, and David Shook, eds. 2015. *Like a New Sun: New Indigenous Mexican*. New York: Phoneme Media.

Tissot, Roger. 2012. "Latin America's Energy Future." Inter-American Dialogue. Working paper. http://www.thedialogue.orgwww.thedialogue.org/PublicationFiles /Tissotpaperweb.pdf.

Torres Cantú, Briceidee. 2016. "La construcción social del riesgo ante proyectos de desarrollo hidro-energéticos en la víspera del fin del mundo. Estudio de tres casos en la Cuenca del Papaloapan y el Istmo de Tehuantepec (1940–2013)." PhD diss., El Colegio de Michoacán.

Tsing, Anna Lowenhaupt. 2012. "Unruly Edges: Mushrooms as Companion Species." *Environmental Humanities* 1 (1): 141–54.

Tsing, Anna Lowenhaupt. 2015. *The Mushroom at the End of the World: On the Possibility of Life in Capitalist Ruins*. Princeton, NJ: Princeton University Press.

Turner, Terence. 1995. "An Indigenous People's Struggle for Socially Equitable and Ecologically Sustainable Production: The Kayapo Revolt Against Extractivism." *Journal of Latin American and Caribbean Anthropology* 1 (1): 98–121.

Turner, Terence, and Vanessa Fajans-Turner. 2006. "Political Innovation and Inter-Ethnic Alliance Kayapo Resistance to the Developmentalist State." *Anthropology Today* 22 (5): 3–10.

Tutino, John. 1980. "Rebelión indígena en Tehuantepec." In *Cuadernos Políticos* 24 (April–June): 89–101.

Van Dooren, Thom. 2014. *Flightways: Life and Loss at the Edge of Extinction*. New York: Columbia University Press.

Villagómez Velázquez, Yanga. 2006. *Política hidroagrícola y cambio agrario en Tehuantepec, Oaxaca*. Mexico: El Colegio de Michoacán.

Viveiros de Castro, Eduardo. 1998. "Cosmological Deixis and Amerindian Perspectivism." *Journal of the Royal Anthropological Institute* 4 (3): 469–88.

Wake, David B., and Vance T. Vredenburg. 2008. "Are We in the Midst of the Sixth Mass Extinction? A View from the World of Amphibians." *PNAS* 105 (S1): 11466–73.

Warman, Arturo. 1993. "The Future of the Isthmus and the Juárez Dam." In *Zapotec Struggles: Histories, Politics, and Representations from Juchitán, Oaxaca*, edited by Miguel A. Bartolomé, Leigh Binford, and Howard Campbell, and Nathaniel Tarn. Washington, DC: Smithsonian Institution Press.

Watts, Laura. 2019. *Energy at the End of the World: An Orkney Islands Saga*. Cambridge, MA: MIT Press.

Watts, Michael. 2015. "Securing Oil: Frontiers, Risk, and Spaces of Accumulated Insecurity." In *Subterranean Estates: Life Worlds of Oil and Gas*, edited by Hannah Appel, Arthur Mason, and Michael Watts, 201–26. Ithaca, NY: Cornell University Press.

West, Paige. 2016. *Dispossession and the Environment: Rhetoric and Inequality in Papua New Guinea*. New York: Columbia University Press.

Weston, Kath. 2017. *Animate Planet: Making Visceral Sense of Living in a High-Tech Ecologically Damaged World*. Durham, NC: Duke University Press.

White, Leslie. 1943. "Energy and the Evolution of Culture." *American Anthropologist* 45 (3): 335–56.

Whitecotton, Joseph W. 1985. *Los zapotecos: Príncipes, sacerdotes y campesinos*. Mexico City: Fondo de Cultura Económica.

Wilhite, Harold. 2005. "Why Energy Needs Anthropology." *Anthropology Today* 21 (3): 1–3.

Winnicott, Donald. 1953. "Transitional Objects and Transitional Phenomena—A Study of the First Not-Me Possession." *International Journal of Psycho-Analysis* 34: 89–97.

Winnicott, Donald. 1964. *The Child, the Family, and the Outside World*. London: Pelican.

Winther, Tanja. 2008. *The Impact of Electricity: Development, Desires and Dilemmas*. Oxford: Berg.

Winthereik, Brit Ross. 2018. "Seeing through Infrastructure: Ethnographies of Health IT, Development Aid, Energy and Big Tech." *STS Encounters-DASTS Working Series* 10 (3): 1–26.

Wolf, Eric R., and Edward C. Hansen. 1967. "Caudillo Politics: A Structural Analysis." *Comparative Studies in Society and History* 9: 168–79.

Wolfe, Cary. 2009. *What Is Posthumanism?* Minneapolis: University of Minnesota Press.

Wolsink, Maarten. 2007. "Wind Power Implementation: The Nature of Public Attitudes: Equity and Fairness Instead of 'Back Yard' Motives." *Renewable and Sustainable Energy Reviews* 11: 1188–1207.

World Bank. 2013. "Latin America: Pioneering Law to Help with Climate Change." April 12. http://www.worldbank.org/en/news/feature/2013/04/12/America-Latina -pionera-en-leyes-sobre-cambio-clim-225-tico.

Wulff, Helena, ed. 2017. *The Anthropologist as Writer: Genres and Contexts in the Twenty-First Century*. New York: Berghan Books.

Yusoff, Kathryn. 2013a. "Geologic Life: Prehistory, Climate, Futures in the Anthropocene." *Environment and Planning D: Society and Space* 31 (5): 779–95.

Yusoff, Kathryn. 2013b. "Insensible Worlds: Postrelational Ethics, Indeterminacy, and the (k)Nots of Relating." *Environment and Planning D: Society and Space* 31 (2): 208–26.

Yusoff, Kathryn. 2016. "Anthropogenesis: Origins and Endings in the Anthropocene." *Theory, Culture, and Society* 33 (2): 3–28. http://dx.doi.org/10.1177/0263276415581021.

Zalasiewicz, Jan. 2012. *The Planet in a Pebble: A Journey into Earth's Deep History*. Oxford: Oxford University Press.

Zendejas, Sergio. 1995. "Appropriating Governmental Reforms: The Ejidos as an Arena of Confrontation and Negotiation." In *Rural Transformations Seen from Below: Regional and Local Perspectives from Western Mexico*, ed. Sergio Zendejas and Pieter de Vries, 23–48. San Diego: University of California, Center for US-Mexican Studies.

Index

autoabastecimiento, 51–52, 55
Avila, Edith, 66–68, 126, 175, 215n29

Barad, Karen, 14, 29, 202n33, 202n39
barotrauma, 159–61. *See also* bats; species
barra. *See* Santa Teresa sandbar
barricade in Álvaro Obregon, 96–98, 117, 125–26, 178–79, 221n6. *See also* resistance to the Mareña wind park
Bateson, Gregory, xiii
bats, 158–61, 219n36, 220n46, 220n49. *See also* species
Beas, Carlos, 121, 213n9, 221n5
Benedict, Ruth, xiii, 201n31
Bennett, Jane, 217n19
bienes comunales and ejidales. *See* communal land ownership
Bií Hioxo park, 99–100
binnizá people: about the, 28–29, 205n9; activism and the, ix, 84, 111–13; history of the, 108, 178, 205n9; the Mareña park and the, 70, 182–83, 187, 189f, 213n16; relationship between ikojts and the, ix, 104–5, 178. *See also* ikojts people; indigenous groups
biolegitimacy, 163, 220n54. *See also* species
biopower, xi–xii
Birdlife International, 153
birds, 39–40, 153–61, 168, 219–20nn42–44, 219n36. *See also* species
Boas, Franz, xiii, 201n31
Boyer, Dominic, xi, 199n9
Butler, Judith, 202n39

caciquismo, xii, 8, 80, 106–8, 189–90. *See also* corruption
Calderón, Felipe, ix, 9, 123–24, 215n32
Caminos y Aeropistas de Oaxaca (CAO), 99
capitalism: development and, 51, 190; green, 21, 111, 190–94, 207n3; rise of, 17–18, 195, 204n52; species and, 139–40, 167–69. *See also* neoliberal development
Capitalocene, 17–18, 204n52. *See also* Anthropocene; Plantationocene
caravan, humanitarian, 187, 222n18
carbon credits, 54, 209n25. *See also* clean-energy legislation; climate change
Castellanos, Jorge, 94, 115–16, 121, 213n14
Castro, Fidel, 108f
CDM, 54, 63, 67, 209n23, 209n26
CEMEX, 52
Center for Human Rights, 121
certified emissions reductions (CER), 54, 209n25. *See also* clean-energy legislation; climate change

Cervecería Cuauhtémoc Moctezuma, 55
CFE, 8, 31, 35, 47–52, 58–59, 146, 188, 214n23
Chakrabarty, Dipesh, 201n29, 203n47, 217n16
Chapman, Andrew, 185–86
Charis Castro, Heliodoro, 3, 90
cinnamon-tailed sparrow, 154–55, 219n36, 220n44
Clean Development Mechanism (CDM), 54, 63, 67, 209n23, 209n26
clean-energy legislation, ix, 8, 54, 144–52, 199n13, 209n25, 209n27, 218n32. *See also* renewable energy; wind parks
climate change: the Anthropocene and, 3, 10–11, 201n29; costs of responses to, 9–10, 39, 106, 112–13, 148, 191–92, 208n16; heterogeneity of humanity and, 197n2, 200n21, 217n16; Mexican response to, ix, xii, 8, 206n28, 209n27; renewable energy as a tool to mitigate, 1–8, 43; species and, 138–40, 150–53. *See also* clean-energy legislation; the environment; renewable energy
"The Climate of History" (Chakrabarty), 201n29
Clipper, 31
CNTE 22, 83–84, 212n9
Coalition of Workers, Peasants, and Students of the Isthmus of Tehuantepec (COCEI), 81, 90, 109–11, 117, 187, 213n11
Coca-Cola, 52, 84, 214n19
COCEI, 81, 90, 109–11, 117, 187, 213n11
Colebrook, Claire, xii
colonialism, xiii, 3, 17–19, 169, 202n42, 203n50, 204n52, 205n15, 218n24. *See also* neocolonialism; neoliberal development
Comaroff, Jean, xiii
Comaroff, John, xiii
Comisión Reguladora de Energía (CRE), 47, 51, 54, 208n19
communal land ownership: about, 47–48, 198n7, 206n24, 215n33, 222n11; agriculture and, 198n7, 208n13, 210n42, 215n33; of the Santa Teresa sandbar, 114, 126, 131–33, 210n36, 213n16, 222n11; of wind parks, 35, 123, 194; wind parks and, 4–5, 46, 67, 71, 104, 210n34, 210n42
"Comuneros Accuse Foreign Companies of Wanting to Take over the Land," 121
CONABIO, 162, 165
CONAGUA, 147–49, 205n16
Constitution, Mexican, 47, 131–34, 148
Convenio 169, 215n35
Convention on International Trade in Endangered Species of Wild Fauna and Flora, 162
Coordinadora Nacional de Trabajadores de la Educación, section 22 (CNTE 22), 83–84, 212n9
corporations. *See* transnational capital; *specific corporations*

Gestión Ambiental Omega (GAO), 149
Ghosh, Amitav, 217n19
global North, 14–15, 195, 217n16. *See also* neoliberal development
Great Acceleration, 3, 18, 102, 212n3. *See also* Anthropocene; fossil fuel industry; trucks
green capitalism, 21, 111, 190–94, 207n3. *See also* capitalism; neoliberal development; wind parks
greenhouse gases, 43, 58–62, 148, 170, 191, 217n16. *See also* fossil fuel industry; oil
Guevara, Che, 106, 108f
Gurrión, Daniel, 58, 80–83
Gutierrez Luis, Beatriz, 166

habitat loss, 17, 138–45, 153–56, 161–66, 220n46, 220n52. *See also* species
Haraway, Donna, 12, 137, 201n30, 202n40
hares, 141–45, 219n36. *See also* species
Harman, Graham, 219n41
Heidegger, Martin, 205n14, 212n19
Heineken, 52, 55
Henestrosa, Davis, 212n16
Holocene, 15, 200n20
Huave people. *See* ikojts people
human rights claims and resistance to the Mareña park, 35, 118–22, 131, 181, 215n32. *See also* resistance to the Mareña wind park
hydrospheres, 161–63, 202n43. *See also* fishing; species

Iberdrola, 48, 75, 79
ICIM, 183
IDB, 30, 60, 83–84, 115, 121, 145, 157–61, 172, 181–83, 207n3, 220n49, 222n15
ikojts people: failure of the Mareña park and the, 178, 182–83, 189f, 213n16; opposition to the Mareña park and the, ix, 65, 70, 110, 122, 213n16; relationship between binnizá and the, ix, 104–5, 178; relationship between species and the, 162, 165, 183. *See also* binnizá people; indigenous groups
imperialism. *See* colonialism; neocolonialism
INAH, 65
inconformes, 105, 113–17, 121–24, 213n2. *See also* resistance to the Mareña wind park
Independent Consultation and Investigation Mechanism (ICIM/MICI), 183
Indian Law Resource Center, 182, 222n10
indicators: machines as, 20, 75–76, 102, 212n19; species as, 75, 102, 141, 160–61, 166. *See also* species; trucks
indigenous groups: activism and, xi, 86, 104–6, 110–12, 117–22, 131–36, 222n10; communal land

holding and, 47–48, 198n7; cosmologies of, 28–29, 203n50; failure of the Mareña wind park and, 20, 62–63, 177–78, 182, 189f; relationship between species and, 160–62, 165–67, 173, 219n38; relationships between, ix, 104–5, 178
individualism, xiii, 195, 212n3
industrial revolution, 18, 203n47, 204n52. *See also* Anthropocene; fossil fuel industry
Ingold, Tim, 205n13, 206n31
Institute of Electrical Studies, 205n16
Interamerican Development Bank (IDB), 30, 60, 83–84, 115, 121, 145, 157–61, 172, 181–83, 207n3, 220n49, 222n15
International Union for the Conservation of Nature (IUCN), 142–44, 159, 217n23, 220n44
investment in wind parks. *See* transnational capital
Irigaray, Luce, 29, 205n14
Isthmus of Tehuantepec: about wind and wind parks in the, viii–ix, 1–2, 7–9, 23–42, 56, 123–24, 207n7, 208n16, 213n7; aeolian politics and futures in the, xi–xii, 191–94; failure of wind park construction in the, 21–22, 170–90; images and maps of the, 7f, 30f, 49, 53, 205nn15–16; interruption of wind park construction in the, 20–21, 103–36; land ownership in the, 47, 198n7; origins of the Mareña park and wind power in the, 19–20, 43–72, 209n24, 209n28; species in the, 21, 141–69, 216n11; trucks in the, 20, 73–102. *See also* Mexico; Oaxaca; Santa Teresa sandbar; *specific towns*
istmo. *See* Isthmus of Tehuantepec
IUCN, 142–44, 159, 217n23, 220n44
Ixtepec, xii, 31, 35, 49, 64, 123, 206n24, 214n23

Jesup North Pacific Expedition, xiii
La Jornada, 121, 181, 187–88
Juárez, Benito, 108–9
Juchitán, xii, 49, 78–83, 95, 105–10, 114–17, 141, 163, 184–88, 222n12

Kirby, Vicki, 74, 211n2
Krusleski, Hannah, 49, 53, 143
Kyoto Protocol, 9

Labor Party (Mexico), 117
LAERFTE, 209n26
land ownership. *See* communal land ownership
Larkin, Brian, 208n16
Latour, Bruno, 201n30, 211nn1–2
Law of Amparo, 134, 215n32. *See also* amparos
Lévi-Strauss, Claude, 201n27

wind: artistic representations of (continued)
204n2; maps of, 30–32, 49, 52–54, 205n16; multiplicity of, x, 18–19, 23–42, 200n24, 209n24;
relationality of, 11–12, 23, 29, 205n13, 206n31;
species and, 21, 36–40, 137–69. *See also* Mareña
Renovables wind park; wind parks
wind parks: community owned, 35, 123, 194;
ecological impacts of, 9–10, 23–24, 38–40,
172–74, 199n15, 206n30, 207n7; failure of,
21–22, 170–90; future of, 188, 191–95; neoliberal
development and, 7, 43, 48–51, 170, 192–94,
205n5, 209n21; origins of Mexican, viii–ix,
1, 8–9, 19–20, 31–33, 43–72, 208n16, 208n19,
209n24; potential benefits to local communities of, 36–38, 46, 59–62, 67–69, 79–82, 210n39,
210n42; resistance to, 20–21, 34–35, 69–70,
83–136, 165–67, 178–87, 214n19, 221n6; species

and, 3, 21, 39–40, 137–69, 218–19nn35–36;
trucks and, 20, 73–102. *See also* Mareña
Renovables wind park
wind power. *See* wind; wind parks
Wolf, Eric, xiii
Wolf, Margery, xiii
World Bank, 60

Yansa Ixtepec project, 35, 123, 194. *See also*
Ixtepec
Yucatecan Constitution of 1841, 130
Yusoff, Kathryn, 16, 150, 200nn20–21, 203n48

Zapata, Emiliano, 96
Zapatista Uprising, 106, 109–10
Zapotec people. *See* binnizá people
Zárate, Luis, 40